Levente Vitos

Computational Quantum Mechanics for Materials Engineers

The EMTO Method and Applications

Springer

Levente Vitos, Professor
Applied Materials Physics
Department of Materials Science
and Engineering
Royal Institute of Technology
Brinellvägen 23
SE-10044 Stockholm
Sweden

British Library Cataloguing in Publication Data
Vitos, Levente
Computational quantum mechanics for materials engineers : the EMTO method and applications. - (Engineering materials and processes)
1. Density functionals 2. Electronic structure - Mathematical models
I. Title
530.1'2
ISBN-13: 9781846289507

Library of Congress Control Number: 2007929990

Engineering Materials and Processes ISSN 1619-0181

ISBN: 978-1-84628-950-7 e-ISBN: 978-1-84628-951-4

Printed on acid-free paper

9 8 7 6 5 4 3 2 1

Springer Science+Business Media
springer.com

To my loved ones

Preface

For centuries, composers and musicians successfully complemented each other. Composers provided humanity with superb compositions, and musicians have been making attempts to vivify the mysterious staves. Great performances, those that managed to create the expected or sometimes unexpected emotions, always captivated the audience and broke fresh ground for new artworks.

The present work belongs to the world of "instrumentalists" and ventures upon creating new perspectives within the field of electronic structure of solid materials. We also have our own great "composers", such as Walter Kohn and Ole Krogh Andersen. Kohn established the Density Functional theory, the most elegant and useful formulation of the many electron problem, and for that he deservedly won the highest award within the scientific world. Andersen created efficient tools to solve the basic Density Functional equations for solids. Today, his theories are widely used in computational materials science. In January 1997, Andersen, looking for experts within the field, contacted Hans Lomholt Skriver's group in Lyngby, and asked then to vivify his latest tool belonging to the third generation muffin-tin methods. Since then, many new incarnations of these methods have come to light. This monograph reveals an implementation of one of the original approaches, namely the Exact Muffin-Tin Orbitals (EMTO) method.

Today theoretical condensed matter physics, besides its fundamental mission to facilitate the understanding of the properties of solid materials at the atomic level, also strives to predict useful quantitative and qualitative data for the development of high-performance materials. Computational quantum mechanics brings an increasing demand for new techniques, which make theoretical investigations more handleable by today's computers. Despite the tremendous developments during the last few decades, an accurate *ab initio* description of substitutional random alloys seemed unreachable. Recently, with the birth of the EMTO method, a distinct step was made towards the modern computational alloy theory by merging the most efficient theories of random alloys with advanced Density Functional techniques. This method allows one to establish a theoretical insight to the electronic structure of complex engineering materials such as the stainless steels. The second part of this monograph is intended to give an up to date overview of the applications of the EMTO method in describing the basic properties of a wide range of important materials.

The present work is addressed, first of all, to graduate students and post-docs who wish to become more familiar with the muffin-tin methods and, in particular, with the EMTO method, its implementation and application to different systems. For these people, the monograph is expected to make up for an extended manual to the EMTO computer code. Scientist moving in the direction of theoretical modeling of material properties might also find the information collected in this book useful. Materials engineers, keen to learn about the latest developments within computational materials science and about their extent and limitations, are also among the readers I have in mind. E-mails or personal contacts from all these readers pointing out the mistakes and shortcomings of the monograph will be received with great gratitude.

It is impossible to list all the people who contributed to the present monograph in one form or another, but I cannot fail to give special thanks to some of them. I would like to express my deepest appreciation to my colleague, Hans Lomholt Skriver, with whom I started the implementation of the EMTO method. A large part of the EMTO computer code is based on his early work and I am very grateful to him for that. It is a pleasure to recognize my profound indebtedness to Ole Krogh Andersen who introduced me into the field of the third generation muffin-tin methods and allowed me to use several of his unpublished notes. I express my truthful thanks to Janos Kollár for supporting me in developing the full charge density technique and merging it with the EMTO method. Börje Johansson is acknowledged for the numerous nice applications we have made together during the years I spent in his group. I wish to thank to all the people with whom I was working in the last fifteen years for the great stimulation in the writing this book and for the common scientific results quoted here. The aid provided by Noémi by reading the manuscript is kindly appreciated. I also acknowledge the financial support from the Swedish Research Council, the Swedish Foundation for Strategic Research, and the Hungarian Scientific Research Fund.

Finally, the continuous encouragement from my parents, my wife and our daughters in the long process of preparing this work is invaluable. My sincere thanks are due to them.

Uppsala,
June 2007

Levente Vitos

Contents

Part III Auxiliary Material

Part I

The Method

1

Basics of Electronic Structure Calculations

Describing the properties of solids from *first-principles* theory implies solving the Schrödinger equations for a huge number of interacting electrons and nuclei. This is an impossible task even for relatively small systems. The first step to overcome this objection is given by the Born–Oppenheimer approximation. It is assumed that on the timescale of nuclear motion, the electronic subsystem is always in its stationary state. Then the motion of the nuclei is solved separately, and this gives rise to the concept of phonons. The remaining set of stationary Schrödinger equations for electrons is still too large for numerical solution. The *Density Functional Theory* [1, 2, 3] offers an elegant reformulation of this problem. Instead of considering many electrons in the external potential of static nuclei, non-interacting electrons in an effective potential are considered. This effective potential is a functional of the total charge density and it incorporates the effect of all the electrons and nuclei. The complexity of the initial problem is hidden in the exchange-correlation part of the potential. Solving the single-electron equations self-consistently, one obtains the equilibrium electron density and the total energy of the system.

Today, the Density Functional theory gives a quantum-mechanical basis for most of the *ab initio* methods used in computational materials science. These methods have made it possible to study complex solid materials of great industrial relevance. The main aims of these applications are the atomic level understanding of the properties and prediction of new data for the development of high-performance materials. In order to accomplish these goals, the numerical methods for solving the single-electron equations should have sufficient accuracy and efficiency. In this chapter, we shall start with a short summary of the Density Functional theory and the Kohn–Sham scheme. We shall briefly review the most important approximations within the Density Functional theory, and the most widely used methods for ordered as well as for disordered systems. Finally, we shall outline the main features of the *Exact Muffin-tin Orbitals* method and give the organization of the rest of the monograph.

1.1 Density Functional Theory

We consider an interacting electron gas moving in an external potential $v_e(\mathbf{r})$. The original theorem of Hohenberg and Kohn [1] states that the ground state of this system is described by the energy functional

$$E_e[n] = F[n] + \int v_e(\mathbf{r})\, n(\mathbf{r}) d\mathbf{r}, \tag{1.1}$$

where the first term is a universal functional of the electron density $n(\mathbf{r})$ and the second term is the interaction energy with the external potential. According to the variational principle, the minimum of $E_e[n]$ is realized for the equilibrium electron density and it equals the total energy of the electronic system. The universal functional is usually represented as

$$F[n] = T_s[n] + E_H[n] + E_{xc}[n], \tag{1.2}$$

where the two "large" contributions are the kinetic energy of non-interacting particles $T_s[n]$ and the Hartree energy $E_H[n]$. The remaining "small" contribution $E_{xc}[n]$ is the so called exchange-correlation functional .

Within the Kohn–Sham scheme [2], the variational principle leads to the effective single-electron Schrödinger equations

$$\left\{-\nabla^2 + v([n];\mathbf{r})\right\} \Psi_j(\mathbf{r}) = \epsilon_j\, \Psi_j(\mathbf{r}). \tag{1.3}$$

Throughout the book atomic Rydberg units are used[1]. The non-interacting Kohn–Sham system is subject to an effective potential

$$v([n];\mathbf{r}) = v_e(\mathbf{r}) + v_H([n];\mathbf{r}) + \mu_{xc}([n];\mathbf{r}). \tag{1.4}$$

Here the second term is the Hartree potential,

$$v_H([n];\mathbf{r}) = 2\int \frac{n(\mathbf{r}')}{|\mathbf{r}-\mathbf{r}'|}\, d\mathbf{r}', \tag{1.5}$$

and the last term is the exchange-correlation potential defined as the functional derivative of $E_{xc}[n]$, *i.e.*

$$\mu_{xc}([n];\mathbf{r}) = \frac{\delta E_{xc}[n]}{\delta n(\mathbf{r})}. \tag{1.6}$$

This latter includes all the electron–electron interactions beyond the Hartree term. The electron density is calculated from the single-electron orbitals according to

$$n(\mathbf{r}) = \sum_{\epsilon_j \le \epsilon_F} |\Psi_j(\mathbf{r})|^2. \tag{1.7}$$

---

[1] In atomic Rydberg units, $\hbar = 2m = e^2/2 = 1$.

In this expression, the summation runs over all the Kohn–Sham states below the Fermi level ϵ_F, which in turn is obtained from the condition

$$N_e = \int n(\mathbf{r})\, d\mathbf{r}, \tag{1.8}$$

where N_e is the number of electrons. The self-consistent solution of Equations (1.3)–(1.8) is used to compute the ground state energy of the electronic system

$$E_e[n] = T_s[n] + \frac{1}{2}\int v_H([n];\mathbf{r})n(\mathbf{r})d\mathbf{r} \;+\; E_{xc}[n] + \int v_e(\mathbf{r})\, n(\mathbf{r})d\mathbf{r}. \tag{1.9}$$

Assuming that the self-consistency of the Kohn–Sham equations (1.3) is achieved, the non-interacting kinetic energy functional may be expressed from the single-electron energies ϵ_j and the self-consistent effective potential as

$$\begin{aligned} T_s[n] &\equiv \sum_{\epsilon_j \le \epsilon_F} \int \Psi_j^*(\mathbf{r})(-\nabla^2)\Psi_j(\mathbf{r})d\mathbf{r} \\ &= \sum_{\epsilon_j \le \epsilon_F} \epsilon_j \;-\; \int n(\mathbf{r})v([n];\mathbf{r})d\mathbf{r}. \end{aligned} \tag{1.10}$$

For electrons moving in the external potential created by the fixed nuclei located on lattice sites $\mathbf{R}$ we have

$$v_e(\mathbf{r}) = -\sum_R \frac{2Z_R}{|\mathbf{r}-\mathbf{R}|}, \tag{1.11}$$

where Z_R are the nuclear charges. Then the total energy of the system formed by electrons and nuclei is obtained from Equation (1.9) plus the nuclear–nuclear repulsion, *viz.*

$$E_{tot} = E_e[n] + \sum_{RR'}{}' \frac{Z_R Z_{R'}}{|\mathbf{R}-\mathbf{R}'|}. \tag{1.12}$$

The prime indicates that the $R = R'$ term is excluded from the sum. Equations (1.3)–(1.8) and (1.11) represent the non-spin polarized Kohn–Sham scheme for electrons from a solid matter. The spin-density functional formalism is obtained by introducing the two spin densities $n^{\uparrow}(\mathbf{r})$ and $n^{\downarrow}(\mathbf{r})$. They are solutions of the Kohn–Sham equation for the spin dependent effective potential. The spin-up and spin-down channels are connected through the spin polarized exchange-correlation potential

$$\mu_{xc}^{\sigma}([n^{\uparrow}, n^{\downarrow}];\mathbf{r}) = \delta E_{xc}[n^{\uparrow}, n^{\downarrow}]/\delta n^{\sigma}(\mathbf{r}) \tag{1.13}$$

($\sigma =\uparrow$ or $\downarrow$) that depends on both spin densities [3].

The accuracy of the Density Functional theory is, in principle, limited only by the employed approximate functionals describing the exchange and correlation energies. The most common approximations for the exchange-correlation term are briefly reviewed in Appendix A. The popular *Local Density Approximation* (LDA) [4, 5, 6, 7] was found to reproduce the ground-state properties of many systems with surprisingly high accuracy. In particular, the bulk properties of 4*d* and 5*d* transition metals, oxides, *etc.*, or the surface properties of metals [9], are very well described within the LDA. However, there are situations where the LDA turned out to be inappropriate even for a qualitative description. The most spectacular failure of it happens in the case of 3*d* transition metals. For instance, LDA predicts incorrect lowest-energy crystal and magnetic structure for pure Fe [10, 11]. During the last two decades several more accurate exchange-correlation density functionals have become available [9, 12, 13, 14, 15, 16, 17]. The most recent gradient level functionals, *e.g.*, the *Generalized Gradient Approximation* (GGA) [8, 9, 15, 16] or the *Local Airy Gas Approximation* (LAG)[17, 18], predict ground-state properties of solids, including that of the 3*d* metals, which are in closer agreement with experiments than the corresponding LDA results.

1.2 Methods for Solving the Kohn–Sham Equations

Developing accurate and at the same time efficient numerical methods for solving the Kohn–Sham equations has been among the biggest challenges within computational materials science. The accuracy of the methods is crucial, *e.g.*, when one searches for the answers given by different approximations used for the exchange-correlation functionals. The full-potential methods have been designed to fulfill this requirement, and provide the exact local density or gradient level description of solid materials [19, 20, 21, 22, 23, 24, 25, 26, 27]. These methods have been applied to calculate the physical properties of ordered compounds, as well as to study defects in these systems. Though, in principle, these techniques give highly accurate results, they are generally very cumbersome and possess several limitations due to various numerical approximations.

The required accuracy for a Kohn–Sham method is always set by the actual property to be computed. For instance, an approximate solution of the Kohn–Sham equations can provide useful information about properties calculated for a fixed crystal structure, whereas quantities involving lattice distortions or structural energy differences require a high level of accuracy. Because of this, often a compromise has been made between accuracy and efficiency, and methods employing certain approximations have been developed. The very expensive computational efforts of full-potential methods are considerably reduced in the pseudopotential methods [28, 29, 30, 31, 32, 33]. In these methods, the deep-lying core states are excluded from consideration, focusing on the valence electrons only. A full-potential description is kept in the interstitial region, where the bonds are located, whereas the true Coulomb-like potential is replaced with a weak pseudopotential in the region near the nuclei. In practice, one often finds that the physical and chemical properties calculated using pseudopotential methods are almost identical with those obtained using all-electron full-potential methods [34]. The high computational speed attainable in pseudopotential calculations makes it possible to perform *ab initio* molecular dynamics [35], that is, to describe the atomic vibrations at high temperatures.

The third important group of Kohn–Sham methods is built around the *muffin-tin approximation* to the effective potential and electron density. This approximation originates from the observation that the exact crystal potential is atomic-like around the lattice sites (where the core states are located) and almost flat between the atoms. Accordingly, within the muffin-tin approximation one substitutes the Kohn–Sham potential by spherically symmetric potentials centered on atoms plus a constant potential in the interstitial region. This family includes the standard Korringa–Kohn–Rostoker (KKR) [36, 37] and screened-KKR [38] methods as well as the methods based on the Atomic Sphere Approximation (ASA) [39, 40, 41, 42, 43]. Due to the involved approximations, the above muffin-tin methods have mostly been restricted to densely packed systems. Some of the deficiencies could be retrenched by the so called Full Charge Density technique [44, 45].

In the early 1990s, a new muffin-tin formalism was introduced by Ole Krogh Andersen and co-workers [46]. Lifting the most significant approximations present in classical muffin-tin related techniques, this approach brings the group of muffin-tin methods back into the heart of the modern Density Functional methods. Keeping the original name, we will refer to the formalism as the *Exact Muffin-tin Orbitals* (EMTO) theory. Here the term "Exact" refers to the fact that, in contrast to former muffin-tin methods, within the EMTO theory the single-electron equations are solved exactly for the *optimized overlapping muffin-tin* (OOMT) potential [47].

Originally, the *Full Charge Density* (FCD) technique [44, 45] was implemented in connection with the Linear Muffin-tin Orbitals (LMTO) method [39, 40]. The FCD-LMTO proved highly promising in the case of close-packed metals [48, 49], but for systems with low crystal symmetry, it had serious shortcomings due to the inappropriate treatment of the kinetic energy term [45]. The accuracy could be maintained only by including overlap and non-spherical corrections to the kinetic energy [40, 45, 50], and as a consequence the FCD-LMTO method became cumbersome. On the other hand, in the EMTO theory the single-electron kinetic energies are calculated exactly for the OOMT potential. Because of this, the EMTO theory is an ideal ground for an accurate FCD total energy technique.

Most of the above full-potential and pseudopotential methods use the Hamiltonian formalism. This means that the electronic spectrum and wave functions are calculated as the eigenvalues and eigenvectors, respectively, of the corresponding Hamiltonian operator. An equivalent way of solving the

Schrödinger equation is to calculate the Green function, which contains all the information about the electronic spectrum of the system. The Green function formalism [51, 52, 53] is more computationally demanding than the Hamiltonian formalism for ordered systems. However, it is suitable for studying disordered systems such as, for instance, impurities in crystals and random alloys, as well as surfaces and interfaces. Many of the muffin-tin methods have been implemented in Green function formalism.

1.3 Chemical and Magnetic Disorder

The main difficulty in the application of Density Functional Theory to real systems is related to the presence of various kinds of disorder. The most common form of disorder is the breakdown of the long range order of the crystal lattice sites. Most real solid materials have a hierarchy of structures beginning with atoms and ascending through various nano or micrometer level crystalline grains. The misaligned single crystals are separated by grain boundaries, stacking faults, interphase boundaries, *etc.* The only way to establish *first-principles* parameters of these polycrystalline systems is to first derive data of microscopic nature and then transform these data to macroscopic quantities by suitable averaging methods based on statistical mechanics [54, 55].

In single crystals the chemical disorder appears as a consequence of the more or less random distribution of the atoms on the lattice sites. Different types of atoms can substitute each other. In systems with spontaneous magnetic ordering at low temperatures, *e.g.* in elemental Fe, a long range magnetic structure is formed. Above the critical temperature the magnetic interaction energy is overtaken by the magnetic entropy contribution, and, as a consequence, the individual magnetic moments of alloy components become randomly oriented, with vanishing total vector moment. This paramagnetic phase is also a disordered phase formed by randomly distributed atomic moments. The temperature induced disorder in the atomic positions, *i.e.* the lattice vibration, is not considered here.

There are various techniques used to describe the energetics of the fully or partially disordered systems. A formally exact approach to this problem is to perform *ab initio* calculations for a chunk of solid material, in which different kinds of atoms, as well as their magnetic moments, are arranged in a configuration similar to the atomic and magnetic structure of the actual disordered solid solution. If one tries to apply this straightforward approach to calculate, for instance, the compositional dependence of some physical properties of a disordered system, one needs to perform numerous calculations for very large systems. Therefore, this supercell technique is very cumbersome, and mainly semi-empirical [56, 57] or approximate order-N methods[2] have adopted it

[2] In an order-N method the computer time grows linearly with the number of atoms in the system.

[58, 59]. For a more efficient approach, the Density Functional methods should be combined with techniques that have been developed in alloy theory over the years, and which are especially designed to deal with disordered systems in a much more efficient way than the brute-force approach described above.

Within the so-called virtual crystal approximation [60, 61, 62] the disordered alloy is modeled by replacing the real system with an equivalent monoatomic system with masses and potentials defined as concentration weighted averages. This simple model suffers from numerous weaknesses [63], *e.g.*, it is unable to describe correctly the bond proportion and volume effects, and its application has been limited to alloys with nearly identical chemical species.

In the cluster expansion formalism, the configurational dependent energy is expanded in terms of the cluster functions [64, 65]. The expansion coefficients are the effective cluster interaction parameters. When all possible clusters are included in the expansion, this formalism gives the exact energy. In practice, however, only a small set of clusters are needed for a reasonably well converged energy. One defines the largest cluster beyond which interactions are ignored. In the structure inversion method by Connolly and Williams [66], the cluster interaction parameters are obtained by fitting the truncated expansion to a set of total energies calculated for ordered structures. Typically, for a binary alloy 20−30 cluster interactions are needed. For multicomponent systems the number of interactions increase rapidly with the number of species, which limits the cluster expansion approach to few-component systems.

The special quasirandom structures method combines the idea of cluster expansion with the use of supercells [67]. Within this approach one constructs special periodic structures that reproduce with high accuracy the most relevant radial correlation functions of an infinite substitutional random alloy. It has been shown that it is possible to construct the special quasirandom structures with as few as 8 atoms per cell. However, so far special quasirandom structures have been designed only for a few selected concentrations on face centered cubic or body centered cubic underlying lattices [67].

The most powerful technique which allows one to treat systems with substitutional disorder is the *Coherent Potential Approximation* (CPA). This approximation was introduced by Soven and Taylor [68, 69], and Györffy [70] has formulated it in the framework of the multiple scattering theory using the Green function formalism. In the past, the CPA was exclusively combined with standard muffin-tin based Kohn−Sham methods [38, 71, 72, 73, 74, 75, 76, 77]. The involved shape approximation for the potential and density limited the application of the CPA to undistorted close-packed systems, *viz.* solids with face centered cubic, body centered cubic or hexagonal close packed underlying crystal lattice. A few years ago, the CPA was restated within the framework of the EMTO theory [78, 79, 80] and most recently, the concept of the CPA was extended to include the short range order effects within the random alloys [81, 82].

1.4 Resumé of the EMTO Method

The implementation and application of the Exact Muffin-tin Orbitals theory combined with the Coherent Potential Approximation and the Full Charge Density technique is the focus of the present work. The Kohn–Sham method created on this platform is referred to as the EMTO method.

The EMTO is a cellular method. The single-electron Schrödinger equations are solved separately within the units defined around the lattice sites. The unit cells are chosen in such a way that they should give a proper description of the local surroundings for every lattice site. For simplicity, we assume that these units are the so called Wigner–Seitz cells or Voronoy polyhedra[3]. The polyhedron for a given center is constructed by bisecting with a plane each line connecting the actual site with another site, and taking the closed region around the lattice site bounded by these planes. The Kohn–Sham potential is approximated by the optimized overlapping muffin-tin wells. In the present implementation, these wells are obtained within the *Spherical Cell Approximation*. Each spherical potential is centered on a lattice site and can spread beyond the boundary of the actual polyhedron. The local Schrödinger equations are solved for these spherical potentials. The local solutions are functions of the energy ϵ. The matching condition between these individual solutions is provided by free electron solutions. This leads to a KKR-type of equation, which selects those energies $\epsilon = \epsilon_j$ for which the Kohn–Sham orbitals $\Psi_j(\mathbf{r})$ exist. For an arbitrary energy, the local solutions join the free electron solutions with a nonzero kink (discontinuity in the first order derivative). Because of this, within the EMTO formalism the KKR equation is also named as the kink cancelation equation. In the case of alloys, the problem of chemical and magnetic disorder is treated via the CPA equations formulated on the EMTO basis. From the output of the self-consistent EMTO calculation the total charge density from Equation (1.7) is constructed. This density is used to compute the total energy (1.12). The energy functional (1.9) is calculated using the FCD technique. The integrals of the Hartree and exchange-correlation energy components are carried out using the *shape function technique*. The interaction energy between remote polyhedra is taken into account through the standard Madelung term, whereas the interaction between cells with overlapping bounding spheres is calculated by the so called *displaced cell technique*.

The most important and prominent feature of the EMTO method is that this approach is suitable for the determination of the energy changes due to anisotropic lattice distortions in ordered systems, in alloys with chemical disorder as well as in alloys with both chemical and magnetic disorder. During recent years, the EMTO has opened new possibilities in the field of computational alloy theory. There is a long list of applications on alloys and compounds, which were not accessible by former CPA related techniques, but

[3] For strongly inhomogeneous potentials a more appropriate division can be used, which reflects the actual atomic sizes.

they are amenable now. This includes several crucial areas from materials engineering such as the elastic, structural and mechanical properties of random alloys of arbitrary compositions. The present monograph gives a complete account of the EMTO formalism and demonstrates its application through several examples.

1.4.1 The Present Monograph

In Chapter 2, we overview the basics of the Exact Muffin-tin Orbitals method. Here we introduce the exact muffin-tin orbitals as a minimal basis set for solving the Kohn–Sham equations (1.3) for the effective potential (1.4). We define the optimized overlapping muffin-tin wells, as the best overlapping muffin-tin approximation to the full-potential. In the second part of Chapter 2, we discuss the Spherical Cell approximation. This is an important approximation needed to establish a well-behaved optimized overlapping muffin-tin potential in the case when the Wigner–Seitz polyhedra are approximated by spherical cells. The equations presented in Chapter 2 form the basis for a self-consistent EMTO calculation.

The slope matrix or the screened structure constant is a central quantity in the muffin-tin formalism. The Dyson equation for the slope matrix and its energy derivatives is delineated in Chapter 3. Here we illustrate how the screening parameters influence the behavior of the slope matrix and establish the range where the screening transformation leads to a slope matrix which has short range and smooth energy dependence. Efficient parameterizations for the energy dependence are given in the second part of Chapter 3.

Chapter 4 deals with the FCD approach. The shape function technique is described in Section 4.1. The FCD method for computing the kinetic and exchange-correlation energies, as well as the Coulomb interactions inside the cell and between Wigner–Seitz cells with non-overlapping and overlapping bounding spheres is given in Section 4.2. At the end of this chapter, the convergence properties of different energy terms are discussed.

In Chapter 5, we briefly review the important features of the Coherent Potential Approximation. The fundamentals of the Exact Muffin-tin Orbitals–Coherent Potential Approximation approach are also presented in Chapter 5. Since many of the equations for the EMTO-CPA formalism can easily be derived from the equations presented Chapter 2, here we give only those where the extension is not straightforward.

In Chapter 6, it is shown how some ground-state properties can be derived from the output of a self-consistent Kohn–Sham calculation. Examples of applications to study these properties in elementary metals, oxides, metallic alloys, including simple metal alloys, Fe-based alloys, austenitic stainless steels, *etc.*, are presented in Chapters 7–9.

Appendix A lists the most widely used approximations for the exchange and correlation functionals. Appendix B contains useful mathematical definitions and relations. This includes the present convention for the spherical and real harmonics, and for the Bessel and Neumann functions. Finally, in Appendix C, we give some useful details about the EMTO self-consistent total energy calculations.

1.4.2 Genuineness

The formalism presented in Sections 2.1 and 3.1 is to a great extent based on the original works by Andersen *et al.* [46, 47, 83, 84, 85, 86]. Section 5.1 gives a brief account of the Coherent Potential Approximation as introduced by Soven and Györffy [68, 70]. The rest of the monograph is based on the works and results published by the author and co-workers during the last fifteen years.

1.4.3 The EMTO Program Package

The full EMTO program package, with Fortran 90 source files, can be obtained upon request (email to Levente.Vitos@fysik.uu.se).

2

Exact Muffin-tin Orbitals Method

In order to reduce the very expensive computational effort of full-potential methods, often a compromise has been made between the accuracy and efficiency, and methods based on approximate single-electron potentials have been developed. The most widely used approach is based on the physically transparent *Muffin-Tin* (MT) approximation. Within this approximation, the effective potential is represented by non-overlapping spherically symmetric potentials around the atomic nuclei and a constant potential in the interstitial region. Although the mathematical formulation of the MT approach is very elegant, it gives a rather poor representation of the exact potential. The so called *Atomic Sphere Approximation* (ASA) [39] substitutes the space by overlapping spherical cells. The total volume of the ASA spheres is equal to the volume of the real space, and thus the region between spheres is completely removed. Because of the large potential spheres, the ASA brings a real improvement to the MT approximation. However, most of the conventional methods based on the ASA potential use a similar approximation for the Schrödinger and Poisson equations [43]. Therefore, with these methods, reasonably accurate results could only be obtained for close-packed systems. In order to increase the accuracy and extend the ASA methods to open systems, different corrections had to be included [39, 40, 44, 45, 48, 50].

In the 1990s, a breakthrough was made by Andersen and co-workers by developing the *Exact Muffin-Tin Orbitals* (EMTO) theory [46, 84, 85, 86]. This theory is an improved screened Koringa–Kohn–Rostoker method [38, 42], in that large overlapping potential spheres can be used for an accurate representation of the exact single-electron potential [84]. The single-electron states are calculated exactly, while the potential can include certain shape approximations, if required. By separating the approaches used for the single-electron states and for the potential, the accuracy can be sustained at a level comparable to that of the full-potential techniques without detracting significantly from the efficiency. In this chapter, we shall review the EMTO theory and introduce a self-consistent implementation of it within the *Spherical Cell Approximation* for the muffin-tin potential.

2.1 The Exact Muffin-tin Orbitals Formalism

The self-consistent solution of the Kohn–Sham equations (1.3), (1.4) and (1.7) involves two main steps. First, the solution of Equation (1.3) for the effective potential (1.4), and second, the solution of the Poisson equation[1] for the total charge density. In this section, we explicate the first problem within the EMTO formalism.

2.1.1 Optimized Overlapping Muffin-tin Wells

Within the overlapping muffin-tin approximation, the effective single-electron potential in Equation (1.4) is approximated by spherical potential wells $v_R(r_R) - v_0$ centered on lattice sites R plus a constant potential v_0, *viz.*

$$v(\mathbf{r}) \approx v_{mt}(\mathbf{r}) \equiv v_0 + \sum_R [v_R(r_R) - v_0]. \tag{2.1}$$

By definition, $v_R(r_R)$ becomes equal to v_0 outside the potential sphere of radius s_R. For simplicity, here and in the following, we suppress the density dependence of the potential. For the vector coordinate, we use the notation $\mathbf{r}_R \equiv r_R \hat{r}_R = \mathbf{r} - \mathbf{R}$, where r_R is the modulus of $\mathbf{r}_R$, and omit the vector notation for index R.

For fixed potential spheres, the spherical and the constant potentials from the right hand side of Equation (2.1) are determined by optimizing the mean of the squared deviation between $v_{mt}(\mathbf{r})$ and $v(\mathbf{r})$, *i.e.* minimizing the

$$F_v[\{v_R\}, v_0] \equiv \int_\Omega \left\{ v(\mathbf{r}) - v_0 - \sum_R [v_R(r_R) - v_0] \right\}^2 d\mathbf{r} \tag{2.2}$$

functional [47]. Here Ω is a region where the potential optimization is performed, *e.g.*, the unit cell. Since F_v is a functional of the spherical potentials, the minimum condition is expressed as

$$\int_\Omega \delta v_R(\mathbf{r}) \frac{\delta F_v[\{v_R\}, v_0]}{\delta v_R(\mathbf{r})} d\mathbf{r} = 0 \quad \text{for any} \quad R, \tag{2.3}$$

where $\delta/\delta v_R(\mathbf{r})$ stands for the functional derivative, and

$$\frac{\partial F_v[\{v_R\}, v_0]}{\partial v_0} = 0. \tag{2.4}$$

The solution of these integro-differential equations gives the optimal $v_R(r_R)$ and v_0, and leads to the so called *optimized overlapping muffin-tin* (OOMT) potential. The reader is referred to Andersen *et al.* [47] for further details about the potential optimization.

[1] The Hartree potential [87] can be found either by direct integration or as the solution of the Poisson equation $\nabla^2 v_H(\mathbf{r}) = -8\pi n(\mathbf{r})$.

In the case of non-overlapping muffin-tins, Equations (2.3) and (2.4) reduce to the spherical average of the full-potential within the potential sphere, *i.e.*

$$v_R(r_R) = \frac{1}{4\pi}\int v(\mathbf{r})d\hat{r}, \text{ for } r_R \leq s_R, \tag{2.5}$$

and to the space average of the full-potential within the s-interstitial region[2], *i.e.*

$$v_0 = \frac{1}{\Omega_{\mathcal{I}^s}}\int_{\mathcal{I}^s} v(\mathbf{r})d\mathbf{r}, \tag{2.6}$$

where $\mathcal{I}^s$ denotes the s-interstitial region and $\Omega_{\mathcal{I}^s} = \Omega - \sum_R \frac{4\pi s_R^3}{3}$ is the volume of the s-interstitial. Note that Equation (2.6) gives the well-known muffin-tin zero.

The overlap between potential spheres may be described in terms of the linear overlap. The linear overlap between two spheres is defined as the relative difference between the sum of the sphere radii and the distance between them, *i.e.*,

$$\omega_{RR'} \equiv \frac{s_R + s_{R'}}{|\mathbf{R}-\mathbf{R}'|} - 1. \tag{2.7}$$

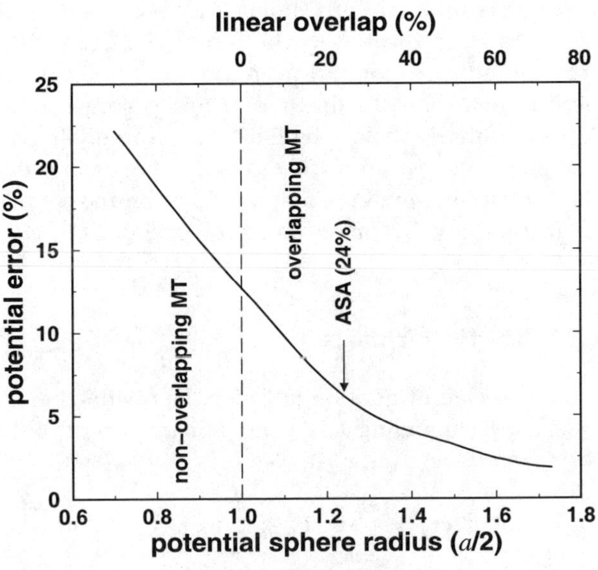

Fig. 2.1. Optimized overlapping muffin-tin potential approximation to the cosine potential in a simple cubic lattice. The radius corresponding to inscribed sphere ($s^i = a/2$) is marked by a vertical line.

---

[2] The interstitial region is the space outside of the potential spheres.

In a monoatomic system, the inscribed sphere is defined as the largest non-overlapping sphere. The radius of this sphere is $s^i_R = \min_{R'} |\mathbf{R} - \mathbf{R}'|/2$. The linear overlap is set to zero for potential spheres with radii $s_R \leq s^i_R$. Obviously, in polyatomic crystals, the inscribed sphere depends upon the convention used to divide the space into units around the lattice sites.

In the following, using a simple model potential, we demonstrate how the full-potential can be represented by overlapping muffin-tins. We model a general three dimensional full-potential by a cosine potential in a simple cubic lattice with lattice constant a. Choosing the reference level in the corner of the Wigner–Seitz cell, *i.e.* in $(a/2, a/2, a/2)$, the cosine potential has the form

$$v_c(\mathbf{r}) = -\cos\left(\frac{2\pi}{a}x\right) - \cos\left(\frac{2\pi}{a}y\right) - \cos\left(\frac{2\pi}{a}z\right) - 3, \qquad (2.8)$$

where x, y, z are the Cartesian coordinates for $\mathbf{r}$. For this potential, we solve Equations (2.3) and (2.4), and the v_0 and $v_R(r_R)$ obtained are used to construct the optimized overlapping muffin-tin potential approximation for $v_c(\mathbf{r})$.

The integrated local deviation between $v_c(\mathbf{r})$ and its OOMT approximation is plotted in Figure 2.1 as a function of the potential sphere radius s. Results are shown from $s = 0.7s^i$ to $s = 1.7s^i$, where $s^i = a/2$ is the radius of the inscribed sphere. This interval corresponds to linear overlaps $\omega_{RR'}$ ranging from -30% to $+70\%$, as indicated at the top of the figure. We observe that the error in the muffin-tin potential decreases continuously with increasing potential sphere radius. Around linear overlaps corresponding to the ASA ($\sim 24\%$), the error falls to below half of the error observed for touching, *i.e.*, non-overlapping spheres. From these results one can clearly see that the accuracy of the overlapping muffin-tin approximation to the full-potential can be improved substantially by increasing the overlap between the potential spheres.

2.1.2 Exact Muffin-tin Orbitals

We solve the single-electron Equation (1.3) for the muffin-tin potential defined in Equation (2.1), by expanding the Kohn–Sham orbital $\Psi_j(\mathbf{r})$ in terms of *exact muffin-tin orbitals* $\bar{\psi}^a_{RL}(\epsilon_j, \mathbf{r}_R)$, *viz.*

$$\Psi_j(\mathbf{r}) = \sum_{RL} \bar{\psi}^a_{RL}(\epsilon_j, \mathbf{r}_R)\, v^a_{RL,j}. \qquad (2.9)$$

The expansion coefficients, $v^a_{RL,j}$, are determined from the condition that the above expansion should be a solution for Equation (1.3) in the entire space. In the EMTO formalism, the algebraic formulation of this matching condition is the so called kink cancelation equation [46, 79, 80]. This equation is equivalent to the Korringa–Kohn–Rostoker tail cancelation equation [36, 37] written in a screened representation [38].

The exact muffin-tin orbitals form a complete basis set for the Kohn–Sham problem. They are defined for each lattice site R and for each $L \equiv (l, m)$, denoting the set of the orbital (l) and magnetic (m) quantum numbers. In practice, it is found that in Equation (2.9), the l summation can be truncated at $l_{max} = 3$, *i.e.* including the s, p, d and f muffin-tin orbitals only.

Screened Spherical Waves

The exact muffin-tin orbitals are constructed using different basis functions inside the potential spheres and in the interstitial region. In the interstitial region, where the potential is approximated by v_0, we use as basis functions the solutions of the wave equation,

$$\left\{\nabla^2 + \kappa^2\right\} \psi^a_{RL}(\kappa^2, \mathbf{r}_R) = 0, \tag{2.10}$$

where $\kappa^2 \equiv \epsilon - v_0$, and ϵ is the energy. Within the EMTO formalism, the $\psi^a_{RL}(\kappa^2, \mathbf{r}_R)$ functions are referred to as the *screened spherical waves* [46].

The boundary conditions for Equation (2.10) are given in conjunction with non-overlapping spheres centered at lattice sites R with radii a_R. Although, the screening sphere radius might also depend upon the orbital quantum number l, for simplicity, here we assume that a_R depends only on the site index R. The screened spherical waves behave like a pure real harmonic $Y_L(\hat{r}_R)$[3] on their own a-spheres, while the $Y_{L'}(\hat{r}_{R'})$ projections on all the other a-spheres, *i.e.* for $R' \neq R$, vanish [46]. With these energy independent boundary conditions, for κ^2 below the bottom of the a-spheres continuum, the screened spherical waves have short range and weak energy dependence [46, 79]. They form a complete basis set in the a-interstitial region and may be expanded in real harmonics $Y_{L'}(\hat{r}_{R'})$ around any site R' as

$$\begin{aligned}\psi^a_{RL}(\kappa^2, \mathbf{r}_R) &= f^a_{Rl}(\kappa^2, r_R)\, Y_L(\hat{r}_R)\delta_{RR'}\,\delta_{LL'} \\ &+ \sum_{L'} g^a_{R'l'}(\kappa^2, r_{R'})\, Y_{L'}(\hat{r}_{R'})\, S^a_{R'L'RL}(\kappa^2).\end{aligned} \tag{2.11}$$

The expansion coefficients, $S^a_{R'L'RL}(\kappa^2)$, are the elements of the *slope matrix*, which is related to the well-known bare KKR structure constant matrix through an inhomogeneous Dyson equation [46]. This is introduced and discussed in Chapter 3.

In Equation (2.11), f^a_{RL} and g^a_{RL} are the value or head and the slope or tail functions, respectively. The previously described boundary conditions for the screened spherical waves for $l \leq l_{max}$ lead to the following conditions at the a-spheres

[3] For the convention used for the real harmonics and for the Bessel and Neumann functions see Appendix B.

$$f^a_{Rl}(\kappa^2, r)|_{a_R} = 1 \text{ and } g^a_{Rl}(\kappa^2, r)|_{a_R} = 0. \tag{2.12}$$

Here we fix the slopes of f^a_{RL} and g^a_{RL} as[4]

$$\left.\frac{\partial f^a_R(\kappa^2, r)}{\partial r}\right|_{a_R} = 0 \text{ and } \left.\frac{\partial g^a_{Rl}(\kappa^2, r)}{\partial r}\right|_{a_R} = \frac{1}{a_R}. \tag{2.13}$$

Using the spherical Bessel and Neumann functions[3], $j_l(\kappa^2, r_R)$ and $n_l(\kappa^2, r_R)$ respectively, the value and slope functions can be expressed as

$$f^a_{Rl}(\kappa^2, r) = t^1_{Rl}(\kappa^2) n_l(\kappa^2, r) + t^2_{Rl}(\kappa^2) j_l(\kappa^2, r) \tag{2.14}$$

and

$$g^a_{Rl}(\kappa^2, r) = -t^3_{Rl}(\kappa^2) n_l(\kappa^2, r) - t^4_{Rl}(\kappa^2) j_l(\kappa^2, r). \tag{2.15}$$

The coefficients $t^{1,\ldots,4}_{Rl}$ are the screening parameters. They are chosen according to the imposed boundary conditions (2.12) and (2.13), namely

$$\left\{\begin{matrix} t^1_{Rl}(\kappa^2) \; t^2_{Rl}(\kappa^2) \\ t^3_{Rl}(\kappa^2) \; t^4_{Rl}(\kappa^2) \end{matrix}\right\} = 2\frac{a_R^2}{w} \left\{\begin{matrix} \frac{\partial j_l(\kappa^2, a_R)}{\partial r_R} & -\frac{\partial n_l(\kappa^2, a_R)}{\partial r_R} \\ \frac{1}{a_R} j_l(\kappa^2, a_R) & -\frac{1}{a_R} n_l(\kappa^2, a_R) \end{matrix}\right\}. \tag{2.16}$$

Here w denotes the average atomic or Wigner–Seitz radius defined from the atomic volume V as

$$\frac{4\pi w^3}{3} \equiv V = \frac{\text{unit cell volume}}{\text{number of atoms in unit cell}}. \tag{2.17}$$

Since the Bessel and Neumann functions satisfy Equation (B.33), for the Wronskian of the value and slope functions we get

$$\mathcal{W}_r\{f^a_{Rl}, g^a_{Rl}\} = a_R, \tag{2.18}$$

and thus the determinant of the screening matrix becomes

$$d^a_{Rl} \equiv t^1_{Rl}(\kappa^2) t^4_{Rl}(\kappa^2) - t^2_{Rl}(\kappa^2) t^3_{Rl}(\kappa^2) = -2\frac{a_R}{w}. \tag{2.19}$$

According to Equations (2.11), (2.14) and (2.15), the screened spherical waves have no pure (lm) character[5], and they are irregular at the origin. This problem is overcome in the next section by replacing the irregular head functions by the partial waves.

---

[4] Different slopes at a-spheres can be used. For example, fixing the slope of g^a to $-1/a^2$ leads to a Hermitian slope matrix.

[5] A function $f_L(\mathbf{r})$ has pure (lm) character if the angular part is fully described by a real harmonic, *viz.*, if $f_L(\mathbf{r}) = f_L(r)\, Y_L(\hat{r})$.

In Equation (2.11), $l \leq l_{max}$ and the l' summation is infinite. In practice, the latter is truncated at $l^h_{max} \approx 8 - 12$. For $l' > l_{max}$, the tail function reduces to the Bessel function, *i.e.* $g^a_{R'l'}(\kappa^2, r_{R'}) = -j_l(\kappa^2, r_{R'})$. These terms are called the *highers* and unlike the low-l components, they are allowed to penetrate into the a-spheres.

Partial Waves
Inside the potential sphere at R, the *partial waves* are chosen as the basis function. These are defined as the products of the regular solutions of the radial Schrödinger equation[6] for the spherical potential $v_R(r_R)$,

$$\frac{\partial^2 \Big[r_R \, \phi_{Rl}(\epsilon, r_R) \Big]}{\partial {r_R}^2} = \left[\frac{l(l+1)}{r_R^2} + v_R(r_R) - \epsilon \right] r_R \, \phi_{Rl}(\epsilon, r_R), \tag{2.20}$$

and the real harmonics, *viz.*

$$\phi^a_{RL}(\epsilon, \mathbf{r}_R) = N^a_{Rl}(\epsilon) \phi_{Rl}(\epsilon, r_R) \, Y_L(\hat{r}_R). \tag{2.21}$$

The normalization function $N^a_{Rl}(\epsilon)$ should be determined from the matching conditions. The partial waves are defined for any real or complex energy ϵ and for $r_R \leq s_R$.

Because a screened spherical wave behaves like $Y_L(\hat{r}_R)$ only on its own a-sphere, the matching condition between $\psi^a_{RL}(\kappa^2, \mathbf{r}_R)$ and $\phi^a_{RL}(\epsilon, \mathbf{r}_R)$ should be set up at this sphere. On the other hand, as we have seen in Section 2.1.1, for an accurate representation of the single-electron potential the potential spheres should overlap. Therefore, usually we have $s_R > a_R$. Because of this, an additional free-electron solution with pure (lm) character has to be introduced. This function realizes the connection between the screened spherical wave at a_R and the partial wave at s_R. It joins continuously and differentiable to the partial wave at s_R and continuously to the screened spherical wave at a_R. Accordingly, the radial part of this backward extrapolated free-electron solution can be written in the form

$$\varphi^a_{Rl}(\epsilon, r_R) = f^a_{Rl}(\kappa^2, r_R) \; + \; g^a_{Rl}(\kappa^2, r_R) \; D^a_{Rl}(\epsilon), \tag{2.22}$$

where $D^a_{Rl}(\epsilon) = \mathcal{D}\{\varphi^a_{Rl}(\epsilon, a_R)\}$ is the logarithmic derivative of $\varphi^a_{Rl}(\epsilon, r_R)$ calculated for $r_R = a_R$. By definition, the logarithmic derivative of a function $f(r_R)$ in the radial mesh point r^0_R is

[6] In practice, we solve the Dirac equation within the so called scalar relativistic approximation rather than the non-relativistic Schrödinger equation. This approximation is obtained by taking into account the mass-velocity and Darwin corrections and neglecting the spin-orbit interaction.

$$\mathcal{D}\{f(r_R^0)\} \equiv \frac{r_R^0}{f(r_R^0)} \frac{\partial f(r_R)}{\partial r_R}\bigg|_{r_R=r_R^0}. \tag{2.23}$$

The normalization function in Equation (2.21) and the logarithmic derivative in Equation (2.22) are determined from the conditions

$$N_{Rl}^a(\epsilon)\phi_{Rl}(\epsilon, s_R) = \varphi_{Rl}^a(\epsilon, s_R), \tag{2.24}$$

and

$$N_{Rl}^a(\epsilon)\left.\frac{\partial \phi_{Rl}(\epsilon, r_R)}{\partial r_R}\right|_{r_R=s_R} = \left.\frac{\partial \varphi_{Rl}^a(\epsilon, r_R)}{\partial r_R}\right|_{r_R=s_R}, \tag{2.25}$$

After simple mathematics, we obtain

$$\frac{1}{N_{Rl}^a(\epsilon)} = \frac{\phi_{Rl}(\epsilon, s_R)}{f_{Rl}^a(\kappa^2, s_R)} \frac{\mathcal{D}\{\phi_{Rl}(\epsilon, s_R)\} - \mathcal{D}\{g_{Rl}^a(\kappa^2, s_R)\}}{\mathcal{D}\{f_{Rl}^a(\kappa^2, s_R)\} - \mathcal{D}\{g_{Rl}^a(\kappa^2, s_R)\}}, \tag{2.26}$$

and

$$D_{Rl}^a(\epsilon) = -\frac{f_{Rl}^a(\kappa^2, s_R)}{g_{Rl}^a(\kappa^2, s_R)} \frac{\mathcal{D}\{\phi_{Rl}(\epsilon, s_R)\} - \mathcal{D}\{f_{Rl}^a(\kappa^2, s_R)\}}{\mathcal{D}\{\phi_{Rl}(\epsilon, s_R)\} - \mathcal{D}\{g_{Rl}^a(\kappa^2, s_R)\}}. \tag{2.27}$$

In these expressions, $\mathcal{D}\{\phi_{Rl}(\epsilon, s_R)\}$, $\mathcal{D}\{f_{Rl}^a(\kappa^2, s_R)\}$ and $\mathcal{D}\{g_{Rl}^a(\kappa^2, s_R)\}$ represent the logarithmic derivatives calculated according to Equation (2.23). Finally, the exact muffin-tin orbitals are constructed as the superposition of the screened spherical waves (2.11), the partial waves (2.21) and the free-electron solution (2.22), *viz.*

$$\begin{aligned}\bar{\psi}_{RL}^a(\epsilon, \mathbf{r}_R) &= \psi_{RL}^a(\kappa^2, \mathbf{r}_R) + N_{Rl}^a(\epsilon)\, \phi_{Rl}(\epsilon, r_R)\, Y_L(\hat{r}_R) \\ &- \varphi_{Rl}^a(\epsilon, r_R)\, Y_L(\hat{r}_R),\end{aligned} \tag{2.28}$$

where the last two terms are truncated outside the s-spheres.

2.1.3 Kink Cancelation Equation

With the exact muffin-tin orbitals defined in Equation (2.28), the trial wave function (2.9) around site R can be expressed as

$$\begin{aligned}\Psi(\mathbf{r}_R) &= \sum_L N_{Rl}^a(\epsilon)\, \phi_{Rl}(\epsilon, r_R)\, Y_L(\hat{r}_R)\, v_{RL}^a \\ &+ \sum_L \left[f_{Rl}^a(\kappa^2, r_R) v_{RL}^a + g_{Rl}^a(\kappa^2, r_R) \sum_{R'L'} S_{RLR'L'}^a(\kappa^2) v_{R'L'}^a \right] Y_L(\hat{r}_R) \\ &- \sum_L \left[f_{Rl}^a(\kappa^2, r_R) + g_{Rl}^a(\kappa^2, r_R) D_{Rl}^a(\epsilon) \right] Y_L(\hat{r}_R)\, v_{RL}^a.\end{aligned} \tag{2.29}$$

Here the index j has been omitted, emphasizing that the expansion may be written for any energy $\epsilon = \kappa^2 + v_0$. We observe that the head functions of the screened spherical wave (2.11) are canceled by the head functions of the free-electron solution (2.22). After rearranging the terms proportional to the tail function, we obtain

$$\Psi(\mathbf{r}_R) = \sum_L N^a_{Rl}(\epsilon)\phi_{Rl}(\epsilon, r_R)\, Y_L(\hat{r}_R) v^a_{RL} + \sum_L g^a_{Rl}(\kappa^2, r_R)\, Y_L(\hat{r}_R) \times \sum_{R'L'} \left[S^a_{RLR'L'}(\kappa^2) - \delta_{R'R}\delta_{L'L}\, D^a_{Rl}(\epsilon)\right]\, v^a_{R'L'}. \quad (2.30)$$

This trial function will be a solution of Equation (1.3) for the muffin-tin potential (2.1), if inside the s-spheres the $l \le l_{max}$ part of the second term from the right hand side of (2.30) vanishes for any r_R. That is, if the $l \le l_{max}$ components of the screened spherical waves, multiplied by the expansion coefficients, are canceled exactly by $\varphi^a_{Rl}(\epsilon, r_R)\, Y_L(\hat{r}_R)\, v^a_{RL,j}$. This is realized if the kink cancelation equation,

$$\sum_{RL} a_{R'} \left[S^a_{R'L'RL}(\kappa_j^2) \;-\; \delta_{R'R}\delta_{L'L}\, D^a_{Rl}(\epsilon_j)\right] v^a_{RL,j} \;=\; 0, \quad (2.31)$$

is satisfied for all R' and $l' \le l_{max}$. Here and in the following $\kappa_j^2 \equiv \epsilon_j - v_0$, and ϵ_j is a Kohn–Sham single-electron energy for which Equation (2.31) has non-trivial solution. The difference between the slope matrix and the logarithmic derivative matrix is called the kink matrix,

$$K^a_{R'L'RL}(\epsilon_j) \;\equiv\; a_{R'}\, S^a_{R'L'RL}(\kappa_j^2) \;-\; \delta_{R'R}\delta_{L'L}\, a_R D^a_{Rl}(\epsilon_j). \quad (2.32)$$

Using the kink-cancelation equation, the wave function inside the potential sphere at R reduces to

$$\Psi_j(\mathbf{r}_R) \;=\; \sum_L N^a_{Rl}(\epsilon_j)\, \phi_{Rl}(\epsilon_j, r_R)\, Y_L(\hat{r}_R)\, v^a_{RL,j} + \sum_{L'}^{l'>l_{max}} g^a_{Rl'}(\kappa_j^2, r_R)\, Y_{L'}(\hat{r}_R) \sum_{R'L} S^a_{RL'R'L}(\kappa_j^2)\, v^a_{R'L,j}. \quad (2.33)$$

Note that the $l' > l_{max}$ components of $\psi^a_{RL}(\kappa_j^2, \mathbf{r}_R)$ are present in the potential spheres. However, due to the $l(l+1)/r_R^2$ centrifugal term in Equation (2.20), the partial waves for large l converge towards the Bessel functions, *i.e.* towards the second term from the right hand side of Equation (2.33).

The solutions of Equation (2.31) are the single-electron energies and wave functions. These solutions can be obtained from the poles of the path operator $g^a_{R'L'RL}(z)$ defined for a complex energy z by

$$\sum_{R''L''} K^a_{R'L'R''L''}(z)\, g^a_{R''L''RL}(z) = \delta_{R'R}\delta_{L'L}. \tag{2.34}$$

In the case of translation symmetry, in Equations (2.31) and (2.34) the site indices run over the atoms in the primitive cell only, and the slope matrix, the kink matrix, and the path operator depend on the Bloch vector **k** from the first Brillouin zone. The **k** and energy dependent slope matrix is obtained from the Bloch sum

$$S^a_{Q'L'QL}(\kappa^2, \mathbf{k}) = \sum_T e^{i\mathbf{kT}} S^a_{Q'L'(Q+T)L}(\kappa^2), \tag{2.35}$$

where Q' and Q denote two sites from the primitive cell, and $\mathbf{T}$ is a translation vector.

2.1.4 Overlap Matrix

The overlap integral of the partial waves within the potential sphere of radius s_R is the norm of $\phi_{Rl}(\epsilon, r_R)$, *viz.*

$$\int \phi^a_{RL}{}^*(\epsilon, \mathbf{r_R})\phi^a_{RL}(\epsilon, \mathbf{r_R})d\mathbf{r}_R = [N^a_{Rl}(\epsilon)]^2 \int_0^{s_R} \phi_{Rl}(\epsilon, r_R)^2 r_R^2 dr_R. \tag{2.36}$$

The radial integral can be calculated using the radial Schrödinger equation (2.20) and the Green's second theorem [46], and it gives

$$\int_0^{s_R} \phi_{Rl}(\epsilon, r_R)^2 r_R^2 dr_R = -s_R \dot{\mathcal{D}}\{\phi_{Rl}(\epsilon, s_R)\}\phi_{Rl}(\epsilon, s_R)^2, \tag{2.37}$$

where the over-dot denotes the energy derivative

$$\dot{\mathcal{D}}\{\phi_{Rl}(\epsilon, s_R)\} \equiv \frac{\partial \mathcal{D}\{\phi_{Rl}(\epsilon, s_R)\}}{\partial \epsilon}. \tag{2.38}$$

The corresponding expression for the free-electron solution (2.22) obtained between the s-sphere and a-sphere, is

$$\begin{aligned}\int_{a_R}^{s_R} \varphi^a_{Rl}(\epsilon, r_R)^2 r_R^2 dr_R &= \int_0^{s_R} \varphi^a_{Rl}(\epsilon, r_R)^2 r_R^2 dr_R - \int_0^{a_R} \varphi^a_{Rl}(\epsilon, r_R)^2 r_R^2 dr_R \\ &= -\, s_R \dot{\mathcal{D}}\{\varphi_{Rl}(\epsilon, s_R)\}\varphi^a_{Rl}(\epsilon, s_R)^2 + a_R \dot{D}^a_{Rl}(\epsilon), \end{aligned} \tag{2.39}$$

where we have taken into account that $\varphi^a_{Rl}(\epsilon, a_R) = 1$. According to the matching conditions (2.24) and (2.25) we have

$$\mathcal{D}\{\varphi_{Rl}(\epsilon, s_R)\} = \mathcal{D}\{\phi_{Rl}(\epsilon, s_R)\} \tag{2.40}$$

for any ϵ, and thus

$$\dot{\mathcal{D}}\{\varphi_{Rl}(\epsilon, s_R)\} = \dot{\mathcal{D}}\{\phi_{Rl}(\epsilon, s_R)\}. \tag{2.41}$$

Therefore, the first term on the right hand side of Equation (2.39) is equal to the overlap integral of the partial waves (2.37).

The energy derivative of the logarithmic derivative function $\dot{D}^a_{Rl}(\epsilon)$ is calculated from Equation (2.27). The energy derivatives of the Bessel and Neumann functions, and their radial derivatives are obtained from the recurrence relations quoted in Appendix B, whereas $\dot{\mathcal{D}}\{\phi_{Rl}(\epsilon, s_R)\}$ is given in Equation (2.37).

The overlap integral of the screened spherical waves over the a-interstitial is obtained in a similar way using the wave equation (2.10) and Green's second theorem [46], and it has the following simple expression

$$\int_{\mathcal{I}^a} \psi^{a\,*}_{R'L'}(\kappa^2, \mathbf{r}_{R'})\psi^a_{RL}(\kappa^2, \mathbf{r}_R)d\mathbf{r} \;=\; a_R\dot{S}^a_{R'L'RL}(\kappa^2), \tag{2.42}$$

where $\mathcal{I}^a$ denotes a-interstitial. Here we recall that only the low-l components of $\psi^a_{RL}(\kappa, \mathbf{r}_R)$ are truncated outside the $\mathcal{I}^a$. The high-l components are present in the whole space, and their contribution to the overlap integral is included in Equation (2.42). The energy derivative of the slope matrix can be calculated from finite differences around κ^2. Alternatively, one may use an analytic expression derived from the unscreened slope matrix. This will be presented in Section 3.1.

We can now establish the overlap matrix for the exact muffin-tin orbitals (2.28) calculated over the whole space. Let us assume, for the moment, that the potential spheres do not overlap and are smaller than the a-spheres. In this situation, we can split the integral $\int \bar{\psi}^*_{R'L'}(\epsilon, \mathbf{r})\, \bar{\psi}_{RL}(\epsilon, \mathbf{r})\, d\mathbf{r}$ into an integral of the partial waves over the potential sphere, an integral of the free-electron solutions over the region between the a-sphere and the potential sphere and an integral of the screened spherical waves over the a-interstitial. Using Equations (2.37), (2.39) and (2.42), we obtain

$$\begin{aligned}\int \bar{\psi}^*_{R'L'}(\epsilon, \mathbf{r})\, \bar{\psi}_{RL}(\epsilon, \mathbf{r})\, d\mathbf{r}& \\ = a_R\dot{S}^a_{R'L'RL}(\kappa^2) \;-\; a_R\dot{D}^a_{Rl}(\epsilon) \;&=\; \dot{K}^a_{R'L'RL}(\epsilon).\end{aligned} \tag{2.43}$$

For overlapping potential spheres, there are other small terms coming from the overlap region. However, these terms are small, and for reasonable small overlaps the above expression remains valid. For more details, the reader is referred to the work by Andersen and co-workers [46, 84].

2.1.5 The Fermi Level

During the iterations for solving self-consistently the Kohn–Sham equations (1.3), the Fermi energy ϵ_F is established from the condition that the total number of states $N(\epsilon_F)$ below the Fermi level should be equal to the number of electrons N_e from the system, *i.e.*

$$N(\epsilon_F) = N_e. \tag{2.44}$$

In practice, $N(\epsilon_F^*)$ is computed for a series of trial ϵ_F^* and the proper ϵ_F is obtained according to Equation (2.44).

Within the present method, we make use of the residue theorem (Appendix B) to find the number of states below the Fermi energy. According to this theorem, the contour integral of the properly normalized path operator gives the total number of states within the contour. Using the overlap matrix of the exact muffin-tin orbitals (2.43) to normalize the path operator, each electronic state will be normalized correctly within the real space. This leads to the following expression for the total number of states below the Fermi level:

$$N(\epsilon_F) = \frac{1}{2\pi i} \oint_{\epsilon_F} G(z)\, dz, \tag{2.45}$$

where

$$G(z) \equiv \sum_{R'L'RL} g^a_{R'L'RL}(z)\, \dot{K}^a_{RLR'L'}(z) - \sum_{RL} \left(\frac{\dot{D}^a_{Rl}(z)}{D^a_{Rl}(z)} - \sum_{\epsilon^D_{Rl}} \frac{1}{z - \epsilon^D_{Rl}} \right), \tag{2.46}$$

with $l, l' \leq l_{max}$. The energy integral in Equation (2.45) is performed on a complex contour that cuts the real axis below the bottom of the valence band and at ϵ_F. It is easy to see that near each pole ϵ_j, the path operator behaves like $\dot{K}^a_{RLR'L'}(\epsilon_j)/(z - \epsilon_j)$, and, therefore, the first term from the right hand side of Equation (2.46) will contribute with 1 to $N(\epsilon_F)$.

Because of the overlap matrix, the $g^a(z)\dot{K}^a(z)$ term may also include the poles of $\dot{D}^a(z)$. Here, we omit the R and l indices for simplicity. Let us denote by z_0 a pole of $\dot{D}^a(z)$. This pole has no physical meaning and should be removed from $N(\epsilon_F)$. Near z_0, both $D^a(z)$ and $\dot{D}^a(z)$ diverge, and thus $g^a(z)\dot{K}^a(z) \to \dot{D}^a(z)/D^a(z)$ for $z \to z_0$. Therefore, subtracting $\dot{D}^a(z)/D^a(z)$ removes the nonphysical pole z_0 of $g^a(z)\dot{K}^a(z)$. In the second step, we have to restore the real poles of $\dot{D}^a(z)/D^a(z)$ due to the zeros of the logarithmic derivative function. We denote by ϵ^D a real energy where $D^a(z)$ vanishes. Expanding $D^a(z)$ near this energy, we have $D^a(z) \approx \dot{D}^a(\epsilon^D)(z - \epsilon^D) + ...$, and thus $\dot{D}^a(z)/D^a(z) \approx 1/(z - \epsilon^D)$. Hence, $\dot{D}^a(z)/D^a(z) - 1/(z - \epsilon^D)$ contains no poles due to the zeros of the logarithmic derivative function. Note that the second term from the right hand side of Equation (2.46) gives no contribution to $N(\epsilon_F)$ if $\dot{D}^a(z)$ is an analytic function of z inside the complex energy contour.

In Equation (2.43) the negligible terms due to the overlap between potential spheres have been omitted [84]. Besides these terms, $N(\epsilon_F)$ in Equation (2.45) gives the exact number of states at the Fermi level for the muffin-tin potential from Equation (2.1).

2.2 Electron Density

The electron density (1.7) is given in terms of Kohn–Sham single-electron wave functions. From the expansion of $\Psi_j(\mathbf{r})$ (Equation (2.9)), a multi-center form for the charge density can be obtained. Although, this multi-center expression gives a highly accurate charge density in the entire space, its application in the Poisson equation or the total energy functional is very cumbersome. Therefore, we seek a more transparent expression which can easily be used to compute the Hartree and exchange-correlation terms in Equations (1.4) and (1.9). To this end, we turn to the one-center expression (2.33). This expression is valid inside the potential spheres. Nevertheless, due to the kink-cancellation equation, the one-center expression remains valid for $r_R > s_R$ as well, if the normalized partial wave $N^a_{Rl}(\epsilon_j)\ \phi_{Rl}(\epsilon_j, r_R)$ is replaced by the backward extrapolated free-electron solution $\varphi^a_{Rl}(\epsilon_j, r_R)$. We use this expression to set up the one-center form for the charge density.

We divide the total density $n(\mathbf{r})$ into components $n_R(\mathbf{r}_R)$ defined inside the Wigner–Seitz cells,[7] *viz.*

$$n(\mathbf{r}) = \sum_R n_R(\mathbf{r}_R). \tag{2.47}$$

Around each lattice site we expand the density components in terms of the real harmonics, *viz.*

$$n_R(\mathbf{r}_R) = \sum_L n_{RL}(r_R) Y_L(\hat{r}_R). \tag{2.48}$$

The partial components $n_{RL}(r_R)$ are radial functions, which are obtained using Equation (2.33), the residue theorem (Appendix B) and the orthogonality condition for the real harmonics (B.3). The final expression can be cast into the following form:

---

[7] In practice, in order to be able to compute the density gradients and eventually the higher order density derivatives, the partial densities should in fact be defined inside a sphere which is slightly larger than the sphere circumscribed to the Wigner–Seitz cell.

$$n_{RL}(r_R) = \frac{1}{2\pi i}\oint_{\epsilon_F} \sum_{L''L'} C_{LL'L''} Z^a_{Rl''}(z,r_R) \times \tilde{g}^a_{RL''L'}(z)\, Z^a_{Rl'}(z,r_R) dz, \tag{2.49}$$

where $C_{LL'L''}$ are the real Gaunt numbers (Appendix B). For the radial functions the following notation has been introduced:

$$Z^a_{Rl}(z,r_R) = \begin{cases} N^a_{Rl}(z)\phi_{Rl}(z,r_R) & \text{if } l \le l_{max} \text{ and } r_R \le s_R \\ \varphi^a_{Rl}(z,r_R) & \text{if } l \le l_{max} \text{ and } r_R > s_R \\ -j_l(\kappa\, r_R) & \text{if } l > l_{max} \text{ for all } r_R \end{cases}. \tag{2.50}$$

Note that in Equation (2.49), the l'' and l' summations include the *highers* as well, i.e. $l'', l' \le l^h_{max}$. The low-l block of the generalized path operator $\tilde{g}^a_{RL'L}(z)$ is given by

$$\tilde{g}^a_{RL'L}(z) = g^a_{RL'RL}(z) + \frac{\delta_{L'L}}{a_R\, \dot{D}^a_{Rl}(z)} \left(\frac{\dot{D}^a_{Rl}(z)}{D^a_{Rl}(z)} - \sum_{\epsilon^D_{Rl}} \frac{1}{z-\epsilon^D_{Rl}} \right), \tag{2.51}$$

with $l, l' \le l_{max}$. The second term from the right hand side of Equation (2.51) is introduced to remove the nonphysical poles of the normalization function $N^a_{Rl}(z)$. The off-diagonal blocks of $\tilde{g}^a_{RL'L}(z)$ are

$$\tilde{g}^a_{RL'L} = \begin{cases} \sum_{R''L''} g^a_{RL'R''L''} a_{R''} S^a_{R''L''RL} & \text{if } l' \le l_{max}, l > l_{max} \\ \sum_{R''L''} S^a_{RL'R''L''} g^a_{R''L''RL} & \text{if } l' > l_{max}, l \le l_{max} \end{cases}. \tag{2.52}$$

Finally, the high-l block is

$$\tilde{g}^a_{RL'L} = \sum_{R''L''R'''L'''} S^a_{RL'R''L''} g^a_{R''L''R'''L'''}\, a_{R'''} S^a_{R'''L'''RL}, \tag{2.53}$$

with $l', l > l_{max}$. For simplicity, in Equations (2.52) and (2.53), the energy dependence has been suppressed. The high-low and low-high blocks of the slope matrix are calculated by the blowing-up technique [83], which will be introduced in Chapter 3.

The charge density computed from Equations (2.48) and (2.49) is normalized within the unit cell, and for reasonably large l^h_{max}, it is continuous at the cell boundaries. Note that because the real Gaunt numbers vanish for $l > l'' + l'$, the partial components of the charge density are nonzero only for $l \le 2l^h_{max}$. For a reasonably high l^h_{max} (10–12), however, the partial components with $l > l^h_{max}$ are very small. Because of this, the l-truncation in Equation (2.48) is usually set to l^h_{max}.

2.3 The Poisson Equation

Equations (2.31), (2.45) and (2.48) constitute the basis of the method used to solve the Schrödinger equation (1.3). In order to perform a self-consistent calculation, one constructs the electron density from the solutions of the kink cancelation equation and calculates the new effective single-electron potential. Within the EMTO formalism, this latter procedure involves two steps [47]. First, we calculate the full-potential from the total charge density (2.48), and second, we construct the optimized overlapping muffin-tin wells (2.1). Due to the l-truncation in the one-center expression from Equation (2.48), the first step is very demanding and inaccurate in the corners of the unit cell. Furthermore, the expression for the effective potential (1.4) involves an integral over the real space. These types of integrals can be performed using, *e.g.*, the shape function technique (Chapter 4). Applying this technique, however, would unnecessarily overcomplicate the self-consistent iterations. In the next section, we show that within the so called Spherical Cell Approximation [80], both of the above problems can be avoided.

2.3.1 Spherical Cell Approximation

To simplify the solution of the Poisson equation, during the self-consistent iterations, we substitute the Wigner–Seitz cell around each lattice site by a spherical cell. The volume of the spherical cell at R, Ω_{w_R}, may be chosen to be equal to the volume of the Wigner–Seitz cell Ω_R centered on the same lattice site, *i.e.*

$$\Omega_{w_R} \equiv \frac{4\pi}{3} w_R^3 = \Omega_R, \tag{2.54}$$

where w_R is the radius of the spherical cell.

Next, we investigate the overlapping muffin-tin approximation to the full-potential in the case of spherical cells. The integrated local deviation between the full-potential and the OOMT potential was introduced and discussed in Section 2.2. It represents a basic measure to establish the accuracy of the muffin-tin potential. But there are other important quantities which should be considered when searching for the best muffin-tin approximation.

The muffin-tin discontinuity is defined as the jump in the muffin-tin potential at the potential sphere boundary, $[v_R(s_R) - v_0]$. The error in the single-electron energies due to the overlap between the s-spheres is given as

$$\begin{aligned} \Delta E_{one} \equiv \sum_{\epsilon_j \leq \epsilon_F} \Delta \epsilon_j &\approx -\frac{\pi}{24} \sum_{RR'} |\mathbf{R} - \mathbf{R}'|^5 \omega_{RR'}^4 \\ &\times [v_R(s_R) - v_0][v_{R'}(s_{R'}) - v_0] n\left((\mathbf{R} - \mathbf{R}')/2\right). \end{aligned} \tag{2.55}$$

This expression has been obtained by the first order perturbation theory [42, 84]. Accordingly, ΔE_{one} is proportional to the average density within the overlap region $n\left((\mathbf{R}-\mathbf{R}')/2\right)$, the square of the muffin-tin discontinuity, and the fourth order power of the linear overlap defined in Equation (2.7). Therefore, if we can keep the muffin-tin discontinuity small, the overlap errors will be negligible, and thus large overlapping potential spheres can be used.

Another important parameter is the constant potential from Equation (2.1) expressed relative to the Fermi level. The screened spherical waves have short range for energies below the bottom of the hard sphere continuum (Chapter 3). Therefore, in order to have localized slope matrices with a smooth energy dependence for energies below $\kappa^2 \approx \epsilon_F - v_0$, one prefers to have v_0 close to ϵ_F.

We make use again of the model potential (2.8) to illustrate how the above parameters depend on the potential sphere radius. For the integral in Equation (2.2) we consider two different domains: (A) the integral is carried out over the real Wigner–Seitz cells, and (B) the Wigner–Seitz cells are substituted by the spherical cells and the OOMT potential is derived for these cells, *i.e.* the integrals are performed within Ω_{w_R}.

In Figure 2.2, we show three sets of results for the integrated local deviation, the muffin-tin discontinuity and the constant potential. The first set (a, solid line) corresponds to the fully optimized overlapping muffin-tins calculated within the Wigner–Seitz cells (domain A). The second set (b, dotted line) is also obtained from the fully optimized overlapping muffin-tins, but this time they are calculated within the spherical cells rather than the Wigner–Seitz cell (domain B). The third case (c, dashed line) corresponds to partially optimized muffin-tins. In this case the spherical potential is fixed to the spherical part of the full-potential according to Equation (2.5) and for this $v(r)$ the constant potential v_0 is optimized within the domain B.

The integrated local deviations are plotted in the upper part of Figure 2.2. In contrast to the monotonously decreasing error obtained in case a (also shown in Figure 2.1), in the second case the error first decreases with s, at $\sim 15\%$ overlap it starts to increase and diverges at larger overlaps. In the third case, for overlapping s-spheres there is a moderate improvement of the muffin-tin approach relative to the non-overlapping situation, but above 30% overlap the integrated local deviation shows no significant dependence on the radius of the potential spheres.

The muffin-tin discontinuity is shown in the middle panel of Figure 2.2. With increasing overlap between the s-spheres, $[v(s)-v_0]$ converges smoothly to zero in case a and it diverges in case b. When v_0 is optimized for $v(r)$ fixed to the spherical part of the full-potential, $[v(s)-v_0]$ approaches zero at small overlaps and remains close to zero up to linear overlaps of 60–70%. Consequently, in the third case the single-electron energies of monoatomic systems are expected to depend negligibly on the overlap between the potential spheres [80].

The constant potential v_0 is plotted in the lower part of Figure 2.2. In case a, v_0 increases with s and reaches the zero potential level at $\sim 60\%$ overlap. When the muffin-tins are fully optimized inside the spherical cell, v_0 decreases

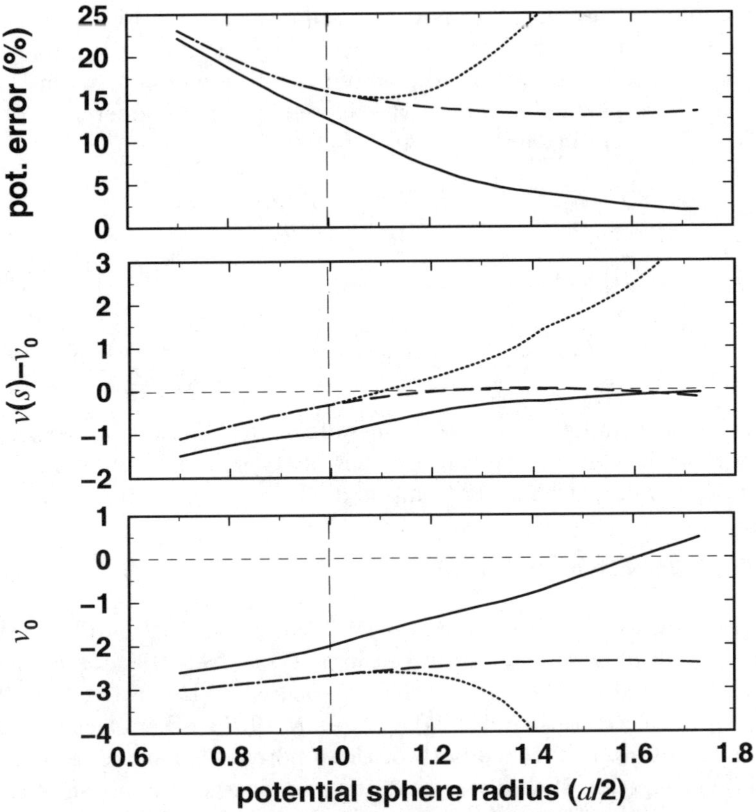

Fig. 2.2. Overlapping muffin-tin potential approximation to the cosine potential in a simple cubic lattice. Upper panel: integrated local deviation of the full-potential and overlapping muffin-tins (in %). Middle panel: muffin-tin discontinuity (in arbitrary units). Lower panel: muffin-tin zero (relative to the zero potential level, in arbitrary units). Solid line: fully optimized overlapping muffin-tins calculated in the Wigner–Seitz cells; dotted line: fully optimized overlapping muffin-tins calculated in the spherical cells; dashed line: spherical potential fixed to the spherical part of the full-potential and muffin-tin zero optimized for this spherical potential.

with s for overlaps larger than $\sim 15\%$. In the third case, v_0 increases slightly with the overlap but always remains well bellow its optimal value, *i.e.* the one from the first case.

From these results one clearly sees that using a spherical cell model, due to the improper description of the full-potential, the fully optimized overlapping muffin-tins approximation breaks down for linear overlaps larger than 10–15%. One possibility to overcome this problem is given by the third case (c), which we will refer to as the Spherical Cell Approximation (SCA).

In summary, the SCA involves two approximations. First, in Equation (2.2) the spherical cells rather than Wigner–Seitz cells are used, and second, $v_R(r_R)$ is fixed to the spherical average of the full-potential given in Equation (2.5). With this particular choice, the expression for parameter v_0 obtained from Equation (2.4) becomes [47, 80]

$$v_0 = \frac{\sum_R \int_{s_R}^{w_R} r_R^2 \left[\int v(\mathbf{r}) d\hat{\mathbf{r}}_R\right] dr_R}{\sum_R \left[4\pi(w_R^3 - s_R^3)/3\right]}. \tag{2.56}$$

When $s_R \to w_R$, the above expression reduces to

$$v_0 \to \frac{\sum_R \left[\int v(\mathbf{r}) d\hat{\mathbf{r}}_R\right]_{r_R = s_R} s_R^2}{\sum_R 4\pi s_R^2} = \frac{\sum_R v_R(s_R) s_R^2}{\sum_R s_R^2}. \tag{2.57}$$

One of the most important consequences of the SCA is that both $v_R(r_R)$ and v_0 from Equation (2.1) are given in terms of the spherical symmetric part of the full-potential, which can be computed efficiently and with high accuracy.

2.3.2 The Effective Potential

In this section, we establish an expression for the spherical part of the single-electron potential (1.4). An electron from a crystal feels the attractive potential (1.11) created by the nuclear charges located at the lattice sites R, and the repulsive electrostatic potential created by all the other electrons, *i.e.* the Hartree potential (1.5). Since both of these potentials have long range, they should be grouped in such a way that at large distance the negative and positive terms cancel each other. Usually, this is done by dividing $v_e(\mathbf{r}) + v_H(\mathbf{r})$ into components due to the charges from inside and from outside of the cell at R. The intra-cell part of the electrostatic potential then becomes

$$v_R^I(\mathbf{r}_R) = -\frac{2\, Z_R}{r_R} + 2 \int_{\Omega_R} \frac{n_R(\mathbf{r}'_R)}{|\mathbf{r}_R - \mathbf{r}'_R|} d\mathbf{r}'_R. \tag{2.58}$$

Using the expansion [87]

$$\frac{1}{|\mathbf{r}_R - \mathbf{r}'_R|} = 4\pi \sum_L \frac{1}{2l+1} \frac{r_R^l}{{r'_R}^{l+1}} Y_L(\hat{r}_R) Y_L(\hat{r}'_R), \tag{2.59}$$

valid for $r'_R > r_R$, we can separate the $\mathbf{r}'_R$ integration in Equation (2.58). Calculating the spherical part of the resulting expression, we arrive at

$$\begin{aligned} v_R^I(r_R) \equiv \frac{1}{4\pi} \int v_R^I(\mathbf{r}_R) d\hat{r}_R &= 8\pi \frac{1}{r_R} \int_0^{r_R} {r'_R}^2 n_{RL_0}(r'_R) dr'_R \\ &+ 8\pi \int_{r_R}^{s_R} r'_R n_{RL_0}(r'_R) dr'_R - \frac{2Z_R}{r_R}, \end{aligned} \tag{2.60}$$

where $n_{RL_0}(r_R)$ is the $L_0 = (0,0)$ partial component of the charge density near site R.

The effect of charges from outside of the potential sphere give the so called Madelung potential

$$v_R^M(\mathbf{r}_R) = -\sum_{R' \neq R} \frac{2\, Z_{R'}}{|\mathbf{r}_{R'} + \mathbf{R}'|} + \sum_{R' \neq R} 2 \int_{\Omega_{R'}} \frac{n_{R'}(\mathbf{r}_{R'})}{|\mathbf{r}_R - \mathbf{r}_{R'} + \mathbf{R} - \mathbf{R}'|} d\mathbf{r}_{R'}. \tag{2.61}$$

This is calculated by expanding $\frac{1}{|\mathbf{r}-\mathbf{r}'|}$ first around $\mathbf{r}_R = \mathbf{r} - \mathbf{R}$ and then around $\mathbf{r}_{R'} = \mathbf{r}' - \mathbf{R}'$, namely

$$\frac{1}{|\mathbf{r} - \mathbf{r}'|} = 4\pi \sum_L \frac{1}{2l+1} \frac{r_R^l}{|\mathbf{r}' - \mathbf{R}|^{l+1}} Y_L(\hat{r}_R) Y_L(\widehat{\mathbf{r}' - \mathbf{R}}), \tag{2.62}$$

and

$$\frac{Y_L(\widehat{\mathbf{r}' - \mathbf{R}})}{|\mathbf{r}' - \mathbf{R}|^{l+1}} = \frac{4\pi}{(2l-1)!!} \sum_{L'L''} C_{LL'L''} \frac{(-1)^{l'}(2l''-1)!!}{(2l'+1)!!} \frac{r_{R'}^l}{|\mathbf{R}' - \mathbf{R}|^{l''+1}} \times Y_{L'}(\hat{r}_{R'}) Y_L(\widehat{\mathbf{R}' - \mathbf{R}}). \tag{2.63}$$

These expansions are strictly valid only for $r_R + r_{R'} < |\mathbf{R}' - \mathbf{R}|$, *i.e.* if the spheres circumscribed on the cells at R and R' do not overlap. The case of overlapping bounding spheres will be discussed in connection with the Madelung energy in Chapter 4. Using the above expressions, for the spherically symmetric part of the Madelung potential we obtain

$$v_R^M \equiv \frac{1}{4\pi} \int v_R^M(\mathbf{r}_R) d\hat{r}_R = \frac{1}{w} \sum_{R' \neq R, L'} M_{RL_0R'L'}\, Q_{R'L'}^{SCA}. \tag{2.64}$$

Here

$$M_{RLR'L'} = 8\pi(-1)^{l'} \sum_{L''} C_{LL'L''} \frac{(2l''-1)!!}{(2l-1)!!(2l'-1)!!} \delta_{l'',l'+l} \times \left(\frac{w}{|\mathbf{R}' - \mathbf{R}|}\right)^{l''+1} Y_{L''}(\widehat{\mathbf{R}' - \mathbf{R}}) \tag{2.65}$$

are the elements of the Madelung matrix, and w is the average atomic radius. Note that because of the Kronecker delta ($\delta_{l'',l'+l}$), in Equation (2.65) only the $l'' = l' + l$ term is nonzero. The multipole moments, Q_{RL}^{SCA}, are calculated within the SCA,

$$Q_{RL}^{SCA} = \frac{\sqrt{4\pi}}{2l+1} \int_0^{w_R} \left(\frac{r_R}{w}\right)^l n_{RL}(r_R)\, r_R^2\, d\, r_R \\ - Z_R\, \delta_{L,L_0} + \delta^{SCA}\, \delta_{L,L_0}. \tag{2.66}$$

Since the integral in Equation (2.66) is performed over the spherical cell rather than over the unit cell, the monopole moments in Equation (2.66) have to be renormalized within the cell [80]. This is realized by the site independent constant δ^{SCA}, which is determined from the condition of charge neutrality, *viz.* $\sum_R Q_{RL_0}^{SCA} = 0$.

Usually, the number of electrons inside the s-sphere,

$$Q(s_R) = \frac{\sqrt{4\pi}}{2l+1} \int_0^{s_R} n_{RL}(r_R)\, r_R^2\, d\, r_R, \tag{2.67}$$

is different from the number of electrons inside the cell, $Q_{RL_0}^{SCA} + Z_R$. This difference contributes a constant shift, Δv_R^M, to the spherical potential. In the SCA, this extra or missing charge is redistributed equally on the N_{NN} nearest-neighbor cells, *i.e.*

$$\Delta v_R^{SCA} = \frac{1}{w} \sum_{R_{NN}} M_{RL_0 R_{NN} L_0}\, \Delta Q_{R_{NN}}, \tag{2.68}$$

where $\Delta Q_{R_{NN}} \equiv \frac{1}{N_{NN}} \left(Q_{RL_0}^{SCA} + Z_R - Q(s_R)\right)$ and R_{NN} are the nearest-neighbor sites.

Finally, the spherical symmetric part of the exchange-correlation potential (1.6) is

$$\mu_{xcR}(r_R) \equiv \frac{1}{4\pi} \int \mu_{xcR}([n_R]; \mathbf{r}_R) d\hat{r}_R. \tag{2.69}$$

The total potential within the potential sphere is obtained as the sum of contributions from Equations (2.60), (2.64), (2.68) and (2.69), namely

$$v_R(r_R) = v_R^I(r_R) + v_R^M + \Delta v_R^{SCA} + \mu_{xcR}(r_R). \tag{2.70}$$

Except the negligible approximations made in the Madelung terms, *i.e.* in Equations (2.64) and (2.68), the above expression gives the exact spherical part of the full potential inside the s-sphere.

In many application, the multipole moments in Equation (2.64) can be neglected for $l > 0$. Moreover, the non-spherical part of $\mu_{xcR}([n_R]; \mathbf{r}_R)$ from the right hand side of Equation (2.69), giving only a small contribution to the spherically symmetric exchange-correlation potential, can also be omitted. In this situation, all potential components in Equation (2.70) depend only on the spherical symmetric density $n_{RL_0}(r_R)$. On the other hand, when the multipole moments for $l \neq 0$ are large (*e.g.*, near the free surfaces), the Madelung potential due to the higher order moments has to be taken into account. In this case, the l-truncation in Equation (2.64) is given by the l-truncation for the multipole moments, which in turn is l_{max}^h used for the charge density (see Section 2.2). In practice, however, the l-truncation for the charge density during the self-consistent iterations can be set to a value (usually 4−6) smaller than the one used to calculate the final full charge density, without sacrificing the accuracy of the method.

Before closing this section, we shall comment on the numerical calculation of the Madelung potential for a periodic bulk (infinite) system. When collecting different contributions to v_R^M, the R' summation in Equation (2.64) should run over all the lattice sites. The Madelung matrix (2.65) can easily be computed for any pair of lattice points R and R'. However, since $M_{RL_0R'L'} \sim 1/|\mathbf{R} - \mathbf{R}'|^{l'+1}$, for the low order multipole moments ($l' \leq 1$), the R' summation diverges. This is because at large distance the number of sites included in the Madelung sum roughly increases as the surface of the coordination shell, *i.e.* as $4\pi|\mathbf{R} - \mathbf{R}'|^2$. On the other hand, the unit cells are neutral, and thus the contribution coming from a remote cell should vanish. In order to overcome this problem, the lattice summation should be carried out using the so called Ewald technique. For details about this technique, the reader is referred to Skriver [39].

2.3.3 Potential Sphere Radius s_R

The potential sphere radius influences the accuracy of the muffin-tin approximation. In Section 2.3.1, we saw that in the case of monoatomic systems the integrated local deviation between full-potential and OOMT potential decreases and the constant potential is pushed towards the Fermi level with increasing potential sphere radius. But what is even more important, large potential spheres also lead to decreased muffin-tin discontinuity and thus to decreased error coming from the s-sphere overlap. On the other hand, this error is proportional to the fourth order power of the linear overlap between spheres [42, 84], which sets an upper limit for s_R.

In Figure 2.3, the overlap error and the muffin-tin discontinuity (inset) are plotted as a function of the potential sphere radius in the case of face centered cubic (fcc) Cu. These calculations were done at the experimental volume using the SCA. We find that for linear overlaps between ~ 4% and ~ 20% the muffin-tin discontinuity is below 10 mRy. This results in a negligible error in the single-electron energies. On the other hand, at larger overlaps the error diverges rapidly with increasing s_R. Taking into account that the integrated local deviation shows a weak s_R dependence for overlaps above ~ 10% (see Figure 2.2), we conclude that in close-packed monoatomic systems the best representation of the full-potential within SCA can be achieved by choosing potential spheres with a linear overlap between 10% and ~ 25%.

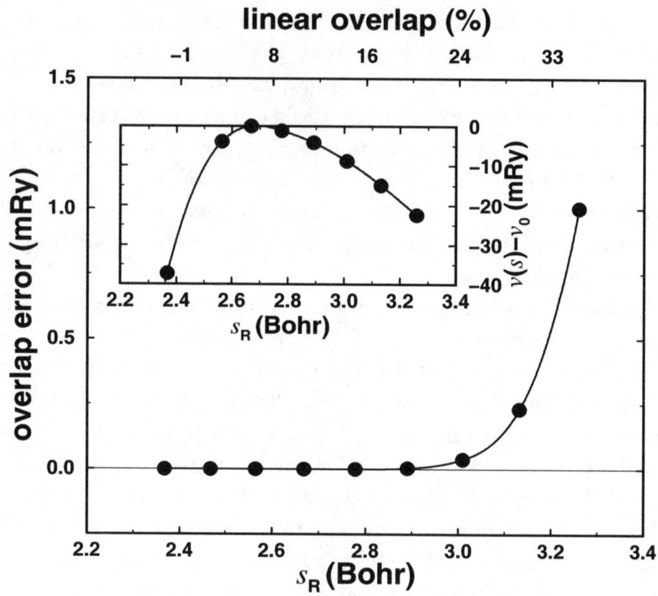

Fig. 2.3. The overlap error for face centered cubic Cu ($w = 2.669$ Bohr) as a function of the potential sphere radius. In the inset the muffin-tin discontinuity is plotted.

In a polyatomic system, small muffin-tin discontinuity can be obtained if the individual spherical potentials at the potential sphere boundary have similar values, *i.e.*

$$v_R(s_R) \approx v_{const.} \quad \text{for each} \quad R. \tag{2.71}$$

Then the constant potential, obtained as the average of the spherical potentials calculated in the vicinity of s_R, will be $v_0 \approx v_{const.}$ and therefore $v_R(s_R) - v_0 \approx 0$. For well localized slope matrices, $v_{const.}$ from Equation (2.71) should have the maximum possible value.

The effect of s_R on the muffin-tin discontinuity in polyatomic crystals is illustrated in the case of magnesium diboride. The crystal structure of MgB_2 has the hexagonal symmetry (space group $P6/mmm$) with $a = 5.833$ Bohr and $c/a = 1.14$ [88, 89, 90]. Layers of Mg and B atoms are located at $z = 0$ and $z = 0.5c/a$, respectively. Using $s_R = w_R$ corresponding to $s_{\text{Mg}}/s_{\text{B}} \approx 1.12$, the linear overlap between nearest-neighbor B atoms is 42%. In this situation, the actual value of the muffin-tin discontinuity is crucial for an accurate self-consistent calculation. In Figure 2.4, we plotted the two muffin-tin discontinuities $[v_{\text{Mg}}(s_{\text{Mg}}) - v_0]$ and $[v_{\text{B}}(s_{\text{B}}) - v_0]$ calculated around Mg and B sites, respectively. Different symbols correspond to different pairs of s_{Mg} and s_{B}. For example, triangles represent muffin-tin discontinuities $+0.288$

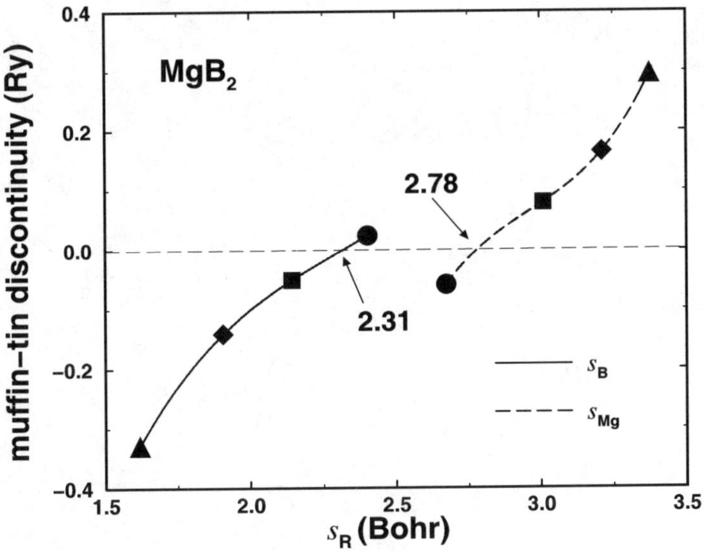

Fig. 2.4. The two muffin-tin discontinuities in MgB_2 compound as functions of the Mg and B potential sphere radii.

Ry and −0.335 Ry obtained using $s_{Mg} = 3.37$ Bohr and $s_B = 1.62$ Bohr. We can see that there is a particular ratio, namely $s_{Mg}/s_B \approx 1.26$, when the muffin-tin discontinuity vanishes on both sites. Although the linear overlap between boron sites is still around 30%, with this choice for s_{Mg} and s_B the error in the single-electron energies is found to be negligible.

3

Slope Matrix

The elements of the slope matrix S^a are the expansion coefficients of the screened spherical waves centered at lattice site R around site R'. They are related to the well-known expansion coefficients of a Neumann function in terms of the Bessel functions. Due to the localization, the slope matrix may be computed on a real space cluster of finite size. Usually, the cluster around R contains the nearest 40−80 lattice sites, depending on the required numerical accuracy. Because of this, the present method, like any screened or tight-binding method, can be applied in real space to treat, *e.g.*, impurities, defects, surfaces, *etc.*

In a self-consistent calculation, we need to know the slope matrix and its energy derivative for a set of complex energies on the energy contour enclosing those states which are considered to be the valence states. During the iterations, the Fermi level is successively updated according to Equation (2.44), and thus the energy points from the complex contour are changed. Because of this, the slope matrix has to be recalculated after each iteration. This is a very cumbersome procedure, especially if the size of the cluster used to compute S^a exceeds 50−70 sites. On the other hand, using a small cluster leads to inaccurate energy derivative of S^a, and thus to large errors in the density of states (2.45) and charge density (2.49). Within the Full Charge Density scheme, which will be introduced in Chapter 4, these errors show up as a loosely normalized charge density and create uncontrollable errors in the total energy. Because of this, one should search for efficient but at the same time highly accurate algorithms to calculate S^a. One possibility is offered by the particularly smooth energy dependence exhibited by the slope matrix, which suggests that accurate parameterized expressions for S^a exist. We shall explore this question within the present chapter.

In Section 3.1, we shall derive the Dyson equation for the slope matrix, and using the low-l ($l', l \leq l_{max}$) block of $S^a_{R'L'RL}$, we shall determine the high-l' and low-l off-diagonal blocks. Furthermore, in this section an analytic expression for the energy derivatives of the slope matrix will be given. We shall investigate the effect of the screening sphere radius (Section 3.2) and real space cluster (Section 3.3) on the localization and energy dependence of S^a. In Section 3.4, we shall discuss the problems related to the numerical calculation of the slope matrix, and present useful representations for an efficient self-consistent implementation.

3.1 Inhomogeneous Dyson Equation

The coefficients $S^a_{R'L'RL}(\kappa^2)$ from expansion (2.11) can be derived from the elements of the bare Korringa–Kohn–Rostoker (KKR) [36, 37] structure constant matrix $S^0_{R'L'RL}(\kappa^2)$. These are defined as the expansion coefficients of the bare spherical wave[1]

$$n_L(\kappa^2, \mathbf{r}_R) \equiv n_l(\kappa^2, r_R)\, Y_L(\hat{r}_R) \tag{3.1}$$

centered on site R, in terms of

$$j_L(\kappa^2, \mathbf{r}_{R'}) \equiv j_l(\kappa^2, r_{R'})\, Y_L(\hat{r}_{R'}) \tag{3.2}$$

centered on site R', *viz.*

$$n_L(\kappa^2, \mathbf{r}_R) = -\sum_{L'} j_{L'}(\kappa^2, \mathbf{r}_{R'})\, S^0_{R'L'RL}(\kappa^2). \tag{3.3}$$

In these expressions, $n_l(\kappa^2, r_R)$ and $j_l(\kappa^2, r_R)$ are the spherical Bessel and Neumann functions, with the conventions given in Appendix B. In Equation (3.3), the expansion coefficients are the KKR structure constants,

$$\begin{aligned} S^0_{R'L'RL}(\kappa^2) &= -8\pi \sum_{L''} C_{LL'L''} \left[-(\kappa w)^2\right]^{\frac{l'+l-l''}{2}} (-1)^l \\ &\times \frac{(2l''-1)!!}{(2l'-1)!!(2l-1)!!}\, n_{L''}(\kappa^2, \mathbf{R}-\mathbf{R}'), \end{aligned} \tag{3.4}$$

where $C_{LL'L''}$ are the real Gaunt numbers (Appendix B). Since the Gaunt numbers vanish unless $l + l + l''$ is even, all the elements of S^0 are real for a real κ^2. This is a consequence of the particular normalization of the Bessel and Neumann functions. We mention that for $\kappa^2 = 0$, the KKR structure constant reduces to $S^0_{R'L'RL}(0) = -M_{R'L'RL}$, where M is the Madelung matrix defined in Equation (2.65).

To find the connection between the slope matrix and the KKR structure constant we start from Equation (2.11), which is the definition of S^a. A screened spherical wave defined in the whole space can be represented by the multi-center expansion

[1] The bare or unscreened representation goes back to the traditional KKR or MTO formalism, where the Neumann (or Hankel) function was chosen as the envelope function in the interstitial region [20, 39, 40].

$$\psi^a_{RL}(\kappa^2, \mathbf{r}) = \sum_{R'L'} \left[f^a_{Rl}(\kappa^2, r_R)\, Y_L(\hat{r}_R) \delta_{RR'}\, \delta_{LL'} \right.$$
$$\left. +\; g^a_{R'l'}(\kappa^2, r_{R'})\, Y_{L'}(\hat{r}_{R'})\, S^a_{R'L'RL}(\kappa^2) \right] . \qquad (3.5)$$

Using the value and slope functions from Equations (2.14) and (2.15), for the above expansion we obtain

$$\psi^a_{RL}(\kappa^2, \mathbf{r}) = \sum_{R'L'} n_{l'}(r_{R'})\, Y_{L'}(\hat{r}_{R'}) \left[t^1_{Rl} \delta_{RR'}\, \delta_{LL'} - t^3_{R'l'} S^a_{R'L'RL} \right]$$
$$- \sum_{R'L'} j_{l'}(r_{R'})\, Y_{L'}(\hat{r}_{R'}) \left[-t^2_{Rl} \delta_{RR'}\, \delta_{LL'} + t^4_{R'l'} S^a_{R'L'RL} \right], \qquad (3.6)$$

where, for simplicity, the common energy argument κ^2 has been dropped. Now, since the Bessel and Neumann functions form a complete basis in the entire space, we should be able to generate the screened spherical wave (3.6) also from the bare spherical waves (3.3). The superposition of the $n_L(\kappa^2, \mathbf{r}_{R'})$ functions leads to

$$\psi^a_{RL}(\kappa^2, \mathbf{r}) = \sum_{R'L'} n_{l'}(r_{R'})\, Y_{L'}(\hat{r}_{R'}) M^a_{R'L'RL}$$
$$- \sum_{R''L''R'L'} j_{l''}(r_{R''})\, Y_{L''}(\hat{r}_{R''}) S^0_{R''L''R'L'} M^a_{R'L'RL}, \qquad (3.7)$$

where M^a is an unknown transformation matrix to be determined. Comparing the coefficients of the Bessel and Neumann functions in Equations (3.6) and (3.7), we find

$$M^a_{R'L'RL} = t^1_{Rl} \delta_{RR'}\, \delta_{LL'} - t^3_{R'l'} S^a_{R'L'RL}, \qquad (3.8)$$

and

$$\sum_{R''L''} S^0_{R'L'R''L''} M^a_{R''L''RL} = -t^2_{Rl} \delta_{RR'}\, \delta_{LL'} + t^4_{R'l'} S^a_{R'L'RL}. \qquad (3.9)$$

After eliminating M^a from Equations (3.8) and (3.9), we obtain

$$- \; t^2_{Rl} \delta_{RR'}\, \delta_{LL'} + t^4_{R'l'} S^a_{R'L'RL}$$
$$= \sum_{R''L''} S^0_{R'L'R''L''} \left[t^1_{Rl} \delta_{R''R'}\, \delta_{L''L'} - t^3_{R''l''} S^a_{R''L''RL} \right] . \qquad (3.10)$$

Rearranging for S^a, we arrive at the inhomogeneous Dyson equation

$$S^a_{R'L'RL} = \frac{t^1_{Rl}}{t^3_{Rl}}\delta_{R'R}\delta_{L'L} + \frac{1}{t^3_{R'l'}}\left[-S^0 - \frac{t^4_{Rl}}{t^3_{Rl}}\right]^{-1}_{R'L'RL}\frac{d^a_{Rl}}{t^3_{Rl}}, \qquad (3.11)$$

where d^a_{Rl} is given in Equation (2.19). Equation (3.11), together with (3.4), is used to compute the slope matrix for an arbitrary complex energy $z = \kappa^2 - v_0$. When the hard sphere radii, a_R, are properly chosen and κ^2 lies below the bottom of the hard sphere continuum, the screened spherical waves have short range, and the slope matrix can be calculated in real space. The optimal choice for the hard sphere radii will be presented in Section 3.2, and the real space cluster used to invert the matrix from the right hand side of Equation (3.11) will be investigated in Section 3.3.

3.1.1 The High–Low Off-diagonal Slope Matrix

In the expressions for the total charge density, Equation (2.49) with (2.52) and (2.53), the elements of the off-diagonal slope matrix with $l^h_{max} \geq l' > l_{max}$ and $l \leq l_{max}$ appear. Computing the full slope matrix up to l^h_{max} would be a time consuming task, not to mention the immense computer capacity needed to store this for a large number of energy points. On the other hand, using the so called blowing-up technique [83], one can generate the high–low off-diagonal blocks of the screened structure constant by using the high–low block of the bare structure constant. We rearrange Equation (3.10) as

$$\begin{aligned} t^4_{R'l'}S^a_{R'L'RL} &= t^2_{Rl}\delta_{R'R}\delta_{L'L} + S^0_{R'L'RL}t^1_{Rl} \\ &- \sum_{R''L''} S^0_{R'L'R''L''}t^3_{R''l''}S^a_{R''L''RL}, \end{aligned} \qquad (3.12)$$

and apply it for $l' > l_{max}$ and $l \leq l_{max}$. According to Equation (2.15), for high-l orbitals, we have $t^3_{Rl} = 0$ and $t^4_{Rl} = 1$. Therefore, the internal summation in Equation (3.12) can be truncated at l_{max}, and the high–low block of the screened slope matrix is obtained from the low–low block as

$$S^a_{R'HRL} = S^0_{R'HRL}t^1_{Rl} - \sum_{R''L''} S^0_{R'HR''L''}t^3_{R''l''}S^a_{R''L''RL}, \qquad (3.13)$$

where $H = (l'm')$ with $l' > l_{max}$. The HL block of the KKR structure constant is obtained from Equation (3.4). The LH block of S^a is found from the HL block (3.13) according to

$$S^a_{R'HRL} = a_R S^a_{RLR'H}. \qquad (3.14)$$

Here we have used the fact that d^aS^a is a Hermitian matrix, as can be seen from Equation (3.11).

3.1.2 Energy Derivatives of S^a

In the present formalism, the energy derivatives of the slope matrix appear in two different places. First, the overlap matrix (2.43) is expressed in terms of the first order energy derivative of S^a. Second, in Section 3.4, we shall demonstrate that the energy dependence of the slope matrix can be represented as an expansion involving the higher order energy derivatives of S^a calculated for a fixed energy. Therefore, we need to develop a systematic algorithm to compute these derivatives. The energy dependence of the slope matrix is usually given in terms of the dimensionless energy parameter $\omega \equiv (\kappa w)^2$, where w is the average atomic radius. After rearranging Equation (3.10), we get

$$\sum_{R''L''} \mathcal{B}_{R'L'R''L''}(\omega)\mathcal{A}_{R''L''RL}(\omega) = -2\frac{a_R}{w}\delta_{R'R}\delta_{L'L}, \tag{3.15}$$

where

$$\mathcal{A}_{R'L'RL}(\omega) \equiv \frac{t^1_{Rl}(\omega)}{t^3_{Rl}(\omega)}\delta_{R'R}\delta_{L'L} - S^a_{R'L'RL}(\omega), \tag{3.16}$$

and

$$\mathcal{B}_{R'L'RL}(\omega) \equiv t^3_{R'l'}(\omega)\left[t^4_{Rl}(\omega)\delta_{R'R}\delta_{L'L} + S^0_{R'L'RL}(\omega)t^3_{Rl}(\omega)\right]. \tag{3.17}$$

Note that the screening parameters, likewise the slope matrix, depend on the energy through ω. Now, applying the product rule,[2] from Equations (3.15) and (3.16) we obtain the nth energy derivative of the slope matrix as

$$\begin{aligned} \frac{d^n S^a(\omega)}{d\omega^n} &= \mathcal{B}(\omega)^{-1}\left[\sum_{i=0}^{n-1} \frac{n!}{(n-i)!i!}\frac{d^{n-i}\mathcal{B}(\omega)}{d\omega^{n-i}}\frac{d^i\mathcal{A}(\omega)}{d\omega^i} + 2\frac{a}{w}\delta_{n,0}\right] \\ &+ \frac{d^n}{d\omega^n}\frac{t^1(\omega)}{t^3(\omega)}, \end{aligned} \tag{3.18}$$

where the RL subscripts have been dropped and the matrix multiplication is implied. The energy derivatives of the bare structure constant $S^0(\omega)$ are calculated directly from Equation (3.4). The derivatives of the screening parameters are obtained from the energy derivatives of the Bessel and Neumann functions presented in Appendix B.

---

[2] The product rule for the nth order derivative of a matrix product NM gives $(NM)^{(n)} = \sum_{i=0}^{n} \frac{n!}{(n-i)!i!} N^{(n-i)}M^{(i)}$, where the superscripts in parenthesis represent the order of the derivative.

3.2 Hard Sphere Radius a_R

The boundary conditions (2.12) and (2.13) lead to specific behavior of the screened spherical waves and slope matrix. First, because on the hard spheres the value and slope functions are energy independent, it is expected that the slope matrix exhibits a smooth and weak energy dependence. Second, since the slope function vanishes on the hard spheres, S^a should decay rapidly with distance.

The degree of localization of the slope matrix may be characterized by the coordination shell radius $d_{R'R} \equiv |\mathbf{R} - \mathbf{R}'|$ dependence of $S^a_{R'L'RL}(\kappa^2)$. Note that for $\omega = 0$, the unscreened structure constants behave as $S^0_{R'L'RL}(0) \sim 1/d^{l'+l+1}_{R'R}$. Any well localized slope matrix should exhibit a significantly faster decay than S^0.

In order to illustrate how S^a and $\dot{S}^a$ depend on the hard sphere radius, we consider the slope matrix of the body centered cubic (*bcc*) lattice. These calculations were carried out for $\omega = 0$, *i.e.*, for a band energy $\epsilon = -v_0$. We used s, p and d orbitals ($l_{max} = 2$), and the real space cluster contained the first 137 nearest-neighbor lattice sites. This includes the site at the origin plus 9 coordination shells with $d_{R'R} \approx 0.87, 1.00, 1.41, 1.66, 1.73, 2.00, 2.18, 2.24$ and

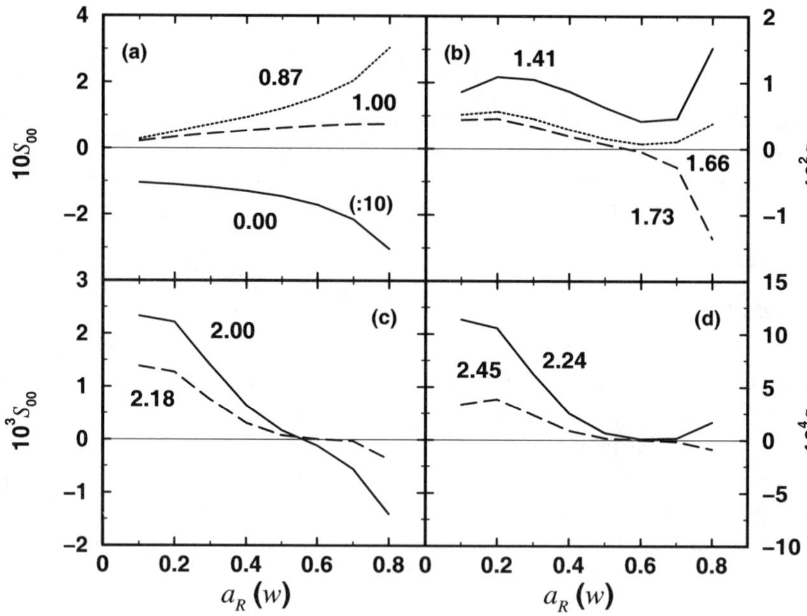

Fig. 3.1. The *ss* element of the *bcc* slope matrix for $\omega = 0$ plotted as a function of the hard sphere radius a_R and the radius of the coordination shell $d_{R'R}$ (shown by numbers in units of lattice constant). Note that in panel (a), the scale for $d_{R'R} = 0.00a$ has been divided by 10.

2.45a, where a is the cubic lattice constant. In the subsequent section, it will be shown that 9 coordination shells in the *bcc* structure are sufficient for a well converged slope matrix in the cluster size.

The L-diagonal elements of $S^a_{R'L'RL}(0)$ are plotted as functions of a_R and $d_{R'R}$ in Figure 3.1. Since the *ss* elements ($L' = L = L_0 \equiv (0,0)$) have the longest range, only these matrix elements are included in the figure. The hard sphere radius is expressed in units of the *bcc* Wigner–Seitz radius $w \approx 0.49a \approx 1.14s^i$, where s^i is the inscribed sphere radius. Results are plotted for $0.1w \leq a_R \leq 0.8w$. Note the scale difference between the four panels corresponding to different groups of $d_{R'R}$.

The worst screening is obtained for $a_R = 0.1w$, where the absolute value of the slope matrix decreases from 1.04 at $d_{R'R} = 0$ (panel (a)) to 2.3×10^{-3} at $d_{R'R} = 2.0a$ (panel (c)) and 3.4×10^{-4} at $d_{R'R} = 2.45a$ (panel (d)). The convergence is somewhat faster for $a_R = 0.7w$, where the on-site term of 2.14 decreases to 5.5×10^{-4} and 1.4×10^{-5} at $d_{R'R} = 2.0a$ and $2.45a$, respectively. With further increase of a_R, the matrix elements for $d_{R'R} \geq 1.41$ start to increase, and the slope matrix becomes less localized. The positive effect of increasing a_R on the localization is even more pronounced in the case of the first order energy derivative $\dot{S}^a_{R'L_0RL_0}(\omega)/d\omega|_{\omega=0}$ shown in Figure 3.2.

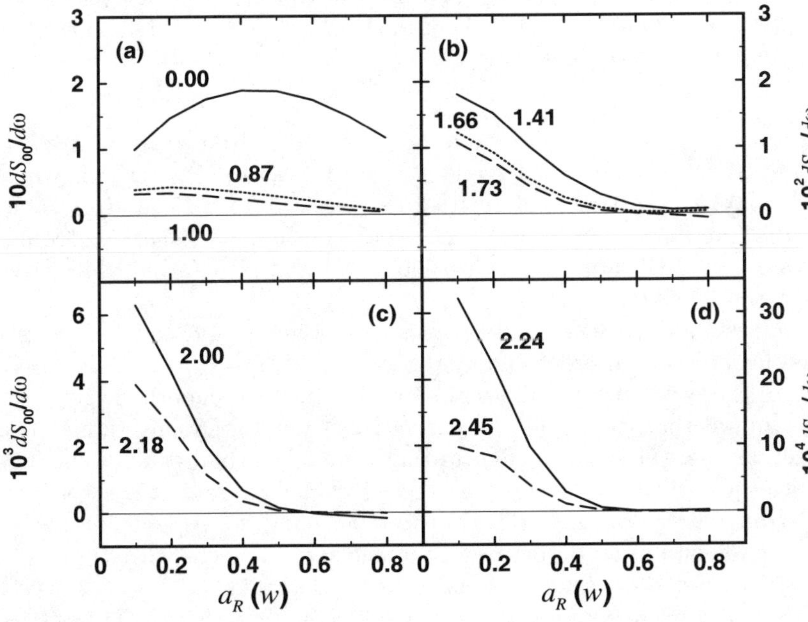

Fig. 3.2. The first order energy derivative of the *ss* element of the *bcc* slope matrix for $\omega = 0$ plotted as a function of the hard sphere radius a_R and the radius of the coordination shell $d_{R'R}$ (shown by numbers in units of lattice constant).

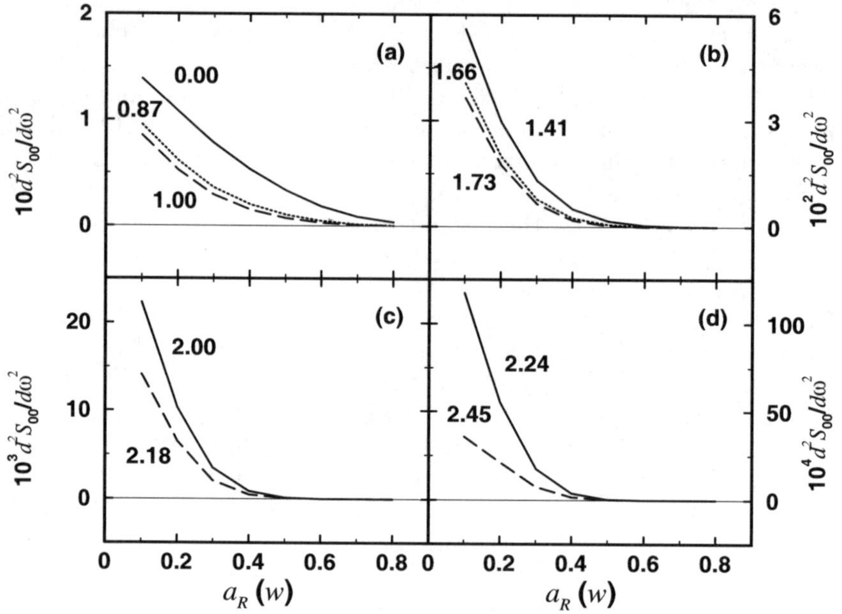

Fig. 3.3. The second order energy derivative of the *ss* element of the *bcc* slope matrix for $\omega = 0$ plotted as a function of the hard sphere radius a_R and the radius of the coordination shell $d_{R'R}$ (shown by numbers in units of lattice constant).

For $a_R = 0.1w$, the absolute value of $\dot{S}^a$ changes from 0.10 at $d_{R'R} = 0$ to 6.3×10^{-3} at $d_{R'R} = 2.0a$ and 9.7×10^{-4} at $d_{R'R} = 2.45a$, whereas for $a_R = 0.7w$, the energy derivative decreases from 0.15 at $d_{R'R} = 0$ to 1.0×10^{-4} at $d_{R'R} = 2.0a$ and 0.2×10^{-5} at $d_{R'R} = 2.45a$. According to Figures 3.1 and 3.2, a good localization of S^a and $\dot{S}^a$ can be achieved with $0.40w \leq a_R \leq 0.75w$.

To investigate the energy dependence, in Figures 3.3 and 3.4, we show the second and fourth order energy derivatives of $S^a_{R'L_0RL_0}(\omega)$ calculated at $\omega = 0$. In these two figures, the scales are identical. A smooth energy dependence requires that the high order derivatives should gradually vanish. It can immediately be seen from the figures that for a_R less than $\approx 0.4w$, the higher order energy derivatives diverge. For these hard sphere radii, the slope matrix has a strong energy dependence. On the other hand, for $a_R > 0.5w$, both the second and fourth order derivatives approach zero for shells with $d_{R'R} \geq 2.00$. For instance, for $a_R = 0.6$, the fourth energy derivative $|d^4 S^a_{R'L_0RL_0}/d\omega^4|$ is 1.6×10^{-3} at $d_{R'R} = 0$ and it drops to 4×10^{-6} at $d_{R'R} = 2.00$. The best convergence of the energy derivatives is obtained for $a_R = 0.8$, where the on-site term of the fourth order derivative is 3.8×10^{-5} and less than 1×10^{-6} for

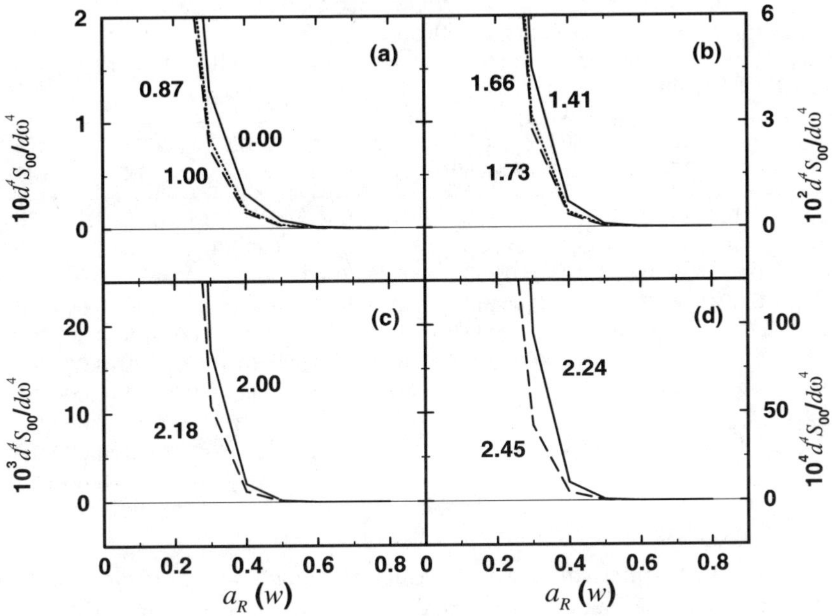

Fig. 3.4. The fourth order energy derivative of the *ss* element of the *bcc* slope matrix for $\omega = 0$ plotted as a function of the hard sphere radius a_R and the radius of the coordination shell $d_{R'R}$ (shown by numbers in units of lattice constant).

$R' \neq R$. Therefore, the *bcc* slope matrix has short range and smooth energy dependence for $0.50w \le a_R \le 0.75w$, corresponding approximately to

$$0.57s^i \le a_R \le 0.85s^i, \tag{3.19}$$

and the best behavior is achieved around $a_R = 0.65w \approx 0.74s^i$. It was found [46], that condition (3.19) approximately holds for any crystal structure.

3.3 Real Space Cluster

For the partial waves explicitly included in the formalism, *i.e.* the low partial waves with $l \le l_{max}$, the hard sphere phase shifts $\alpha_{Rl}(\kappa^2)$ are given by

$$\cot \alpha_{Rl}(\kappa^2) \;=\; -n_l(\kappa^2, a_{Rl})/j_l(\kappa^2, a_{Rl}). \tag{3.20}$$

For the remaining Rl-channels, $\alpha_{Rl}(\kappa^2)$ are the proper phase shifts. For high ls the latter vanish, and at that point the matrix to be inverted in Equation (3.11) can be truncated. In practice, the l-block of the matrix is truncated at l_{max}.

Table 3.1. The *ss* elements of the *bcc* slope matrix calculated at $\omega = 0$ for different cluster sizes. $d_{R'R}$ is the radius of the cluster, and N_s is the number of sites in the cluster. S^a_{0i} is the slope matrix for the ith coordination shell (1.42 is the 3rd shell, 1.66 is the 4th shell and so on).

$d_{R'R}$	1.42	1.66	1.73	2.00	2.18	2.24	2.45
N_s	27	51	59	65	89	113	137
S^a_{00}	-2.146048	-2.146081	-2.146182	-2.146185	-2.146185	-2.146185	-2.146185
S^a_{01}	0.205731	0.205399	0.206237	0.206240	0.206240	0.206240	0.206240
S^a_{02}	0.074536	0.073510	0.073561	0.073622	0.073622	0.073621	0.073621
S^a_{03}	0.004867	0.004228	0.004583	0.004585	0.004594	0.004590	0.004590
S^a_{04}	-	0.000705	0.000999	0.001072	0.001075	0.001066	0.001071
S^a_{05}	-	-	-0.002855	-0.002835	-0.002801	-0.002801	-0.002799
S^a_{06}	-	-	-	-0.000560	-0.000550	-0.000554	-0.000552
S^a_{07}	-	-	-	-	-0.000033	-0.000035	-0.000031
S^a_{08}	-	-	-	-	-	0.000019	0.000018
S^a_{09}	-	-	-	-	-	-	-0.000014

The screened spherical waves have short range, which means that the slope matrix can be calculated in real space. For each site R, we set up a finite cluster formed by the first few nearest-neighbor lattice sites, and invert the matrix from the right hand side of Equation (3.11) on that cluster. For positive energies the screened spherical waves for a finite cluster exhibit surface resonances. To cure this problem a concave hard sphere, so called Watson sphere, should be included, which encloses the cluster of hard spheres. The radius of this sphere is chosen to be somewhat larger than the radius of the cluster plus the largest a_R. The maximum orbital quantum number on the Watson sphere is l^w_{max}. Usually, l^w_{max} =6−8 is sufficient for positive energies $\omega \leq 5$.

We investigate the convergence of the slope matrix in terms of the cluster size by calculating the *bcc* slope matrix for clusters consisting of $N_s = 27, 51, 59, 65, 89, 113$ and 137 nearest-neighbor sites. These calculations were performed for $l_{max} = 2$, $l^w_{max} = 8$, using a hard sphere radius of $a_R = 0.70w \approx 0.80s^i$. In the previous section, we have seen that this choice leads to well localized screened spherical waves. The energy was fixed at $\omega = 0$.

The *ss* elements of the *bcc* slope matrix are listed in Table 3.1. First of all, we can see that clusters containing more than 65 sites lead to almost vanishing slope matrix elements at the clusters boundaries. For clusters of $27, 51$ and 59 sites, there is still a sizable contribution coming from the farthest coordination shell. Because of this, small clusters are not suitable to produce a well converged slope matrix. It is only for clusters larger than 65 atoms, where the matrix elements remain almost unchanged with increasing N_s. A similar conclusion can be drawn from the cluster size dependence of the first energy derivative of the slope matrix, listed in Table 3.2.

Table 3.2. The first order energy derivative of the *ss* elements of the *bcc* slope matrix calculated at $\omega = 0$ for different cluster sizes. For notation see caption for Table 3.1.

$d_{R'R}$	1.42	1.66	1.73	2.00	2.18	2.24	2.45
N_s	27	51	59	65	89	113	137
$\dot{S}^a_{00}$	0.147962	0.147951	0.147925	0.147923	0.147923	0.147923	0.147923
$\dot{S}^a_{01}$	0.013977	0.013861	0.014056	0.014057	0.014057	0.014057	0.014057
$\dot{S}^a_{02}$	0.007630	0.007332	0.007347	0.007368	0.007368	0.007368	0.007368
$\dot{S}^a_{03}$	0.000679	0.000479	0.000586	0.000587	0.000590	0.000588	0.000588
$\dot{S}^a_{04}$	-	0.000052	0.000137	0.000163	0.000164	0.000161	0.000162
$\dot{S}^a_{05}$	-	-	-0.000256	-0.000259	-0.000250	-0.000250	-0.000249
$\dot{S}^a_{06}$	-	-	-	-0.000105	-0.000103	-0.000105	-0.000104
$\dot{S}^a_{07}$	-	-	-	-	-0.000003	-0.000004	-0.000002
$\dot{S}^a_{08}$	-	-	-	-	-	0.000003	0.000003
$\dot{S}^a_{09}$	-	-	-	-	-	-	-0.000002

It has been found that for most of the crystal structures, clusters of 60−90 sites are sufficient to obtain slope matrices with high accuracy. Of course, the actual size depends on the screening sphere radii. For an optimal set of a_R and a large l^w_{max} the cluster size may be decreased further without loosing significantly from the accuracy.

3.4 Numerical Determination of the Slope Matrix

In the previous section, we have seen that with properly chosen hard sphere radii, the high order energy derivatives of the EMTO slope matrix are negligible. This suggests that a polynomial expansion of $S^a(\omega)$ around ω_0 might be used to calculate the slope matrix for an arbitrary real or complex energy. In the following, we discuss separately systems which have a small characteristic valence bandwidth ($\lesssim 1.0$ Ry) and systems with a large bandwidth.

3.4.1 Systems with Narrow Bandwidth

For a self-consistent calculation, the slope matrix is required within an energy window that includes the valence band plus ~ 0.2 Ry above the Fermi level, *i.e.*, for energies $\epsilon_b \leq \epsilon \leq \epsilon_F + 0.2$ Ry, where ϵ_b is an energy slightly below the bottom of the valence band. Energies above the Fermi level are needed when searching for the Fermi level for the next iteration. In systems with deep-lying core states and a narrow valence band structure, we can take $\epsilon_F - \epsilon_b \approx 1.4$ Ry. This is the case for most of the middle and late transition metals, simple metals, *etc.*, and their alloys at ambient conditions. Since $\epsilon_F - v_0$ is typically around 0.6 ± 0.2 Ry, for such systems the slope matrix should be calculated for -0.8 Ry $\leq \omega/w^2 \leq 0.8$ Ry. Using an average $w^2 = 10$ Bohr^2, the energy interval of interest turns out to be ± 8 around $\omega = 0$. For these energies, we expand the slope matrix in Taylor series around ω_0, *viz.*

$$S^a(\omega) = S^a(\omega_0) + \frac{1}{1!}\frac{dS^a(\omega)}{d\omega}(\omega - \omega_0) + \frac{1}{2!}\frac{d^2S^a(\omega)}{d\omega^2}(\omega - \omega_0)^2 + ... \quad (3.21)$$

where, for simplicity, we have dropped the RL subscripts. The first and higher order energy derivatives are computed using the analytic expression (3.18). In practice, the expansion center ω_0 is chosen somewhere close to 0.

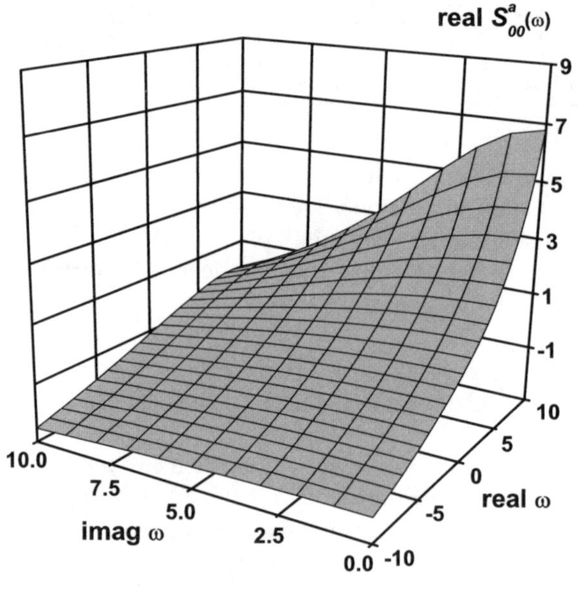

Fig. 3.5. Real part of the ss element of the slope matrix $S^a(\omega)$ calculated for the fcc structure.

To test the accuracy of the above Taylor expansion we consider the slope matrix in the complex energy plane, where $S^a(\omega)$ will have both real and imaginary components. In Figures 3.5 and 3.6, we plotted the real and imaginary parts of the ss element of the fcc slope matrix (*i.e.* elements with $L = L' = (0, 0)$) as a function of ω for an interval corresponding to $-10 \leq Re(\omega) \leq 10$ and $0 \leq Im(\omega) \leq 10$. Since

$$Re[S^a(x + iy)] = Re[S^a(x - iy)], \quad (3.22)$$

and

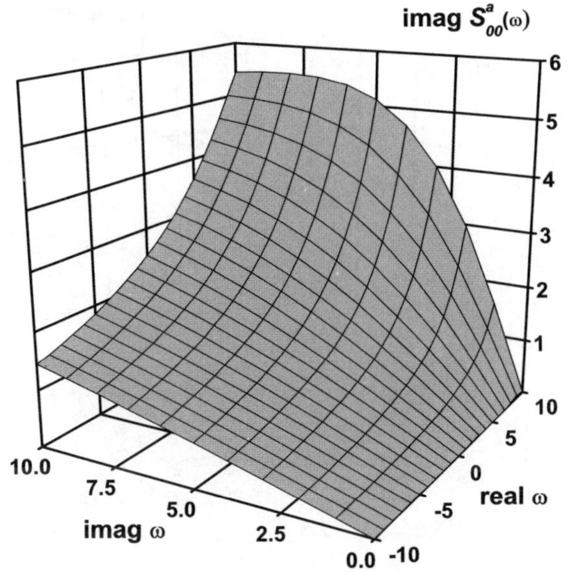

Fig. 3.6. Imaginary part of the ss element of the slope matrix $S^a(\omega)$ calculated for the fcc structure.

$$Im[S^a(x+iy)] = -Im[S^a(x-iy)], \tag{3.23}$$

where $x = Re(\omega)$ and $y = Im(\omega)$, in figures results are shown only for positive $Im(\omega)$. The reason for showing only the ss sub-block is that this is the largest and the most delocalized one within the real space. The fcc slope matrix was calculated on a cluster containing the 79 nearest-neighbor lattice sites using s, p and d orbitals and $a_R \approx 0.77s^i$. The real space slope matrix was transformed to the reciprocal space by means of Equation (2.35). In figures, the slope matrix is shown for the $\mathbf{k} = (0,0,0)$ point from the Brillouin zone. These figures confirm the smooth energy dependence of the slope matrix in the complex energy plane. Note that the imaginary part of $S^a(\omega)$ vanishes for $Im(\omega) = 0$.

The accuracy of the Taylor expansion (3.21) for the slope matrix can be established by computing the relative deviation between the exact value (S_{exact}) from Figures 3.5 and 3.6 and those calculated using the expansion (S_{expan}). The relative error, defined as [91]

$$\frac{\{[Re(S_{exact} - S_{expan})]^2 + [Im(S_{exact} - S_{expan})]^2\}^{1/2}}{\{[Re(S_{exact})]^2 + [Im(S_{exact})]^2\}^{1/2}}, \tag{3.24}$$

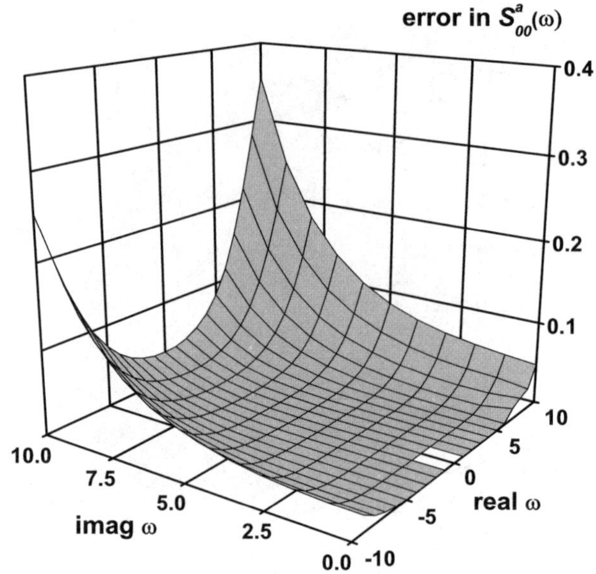

Fig. 3.7. The relative error of expansion (3.21) for the *ss* element of the *fcc* slope matrix using six derivatives. The expansion center is $\omega_0 = (0, 0)$.

is plotted in Figure 3.7 as a function of ω for $\omega_0 = (0, 0)$. We observe a gradually increasing error with increasing $|\omega|$. The expansion reproduces the calculated slope matrix with an accuracy better than $\sim 1\%$ within a radius of ~ 5 around the expansion center. For a radius of ~ 10 the accuracy of the expansion is still below 5%. Therefore, for narrow bands (*i.e.* for $|\omega| \leq 8$) a 6th order Taylor expansion around $\omega_0 = (0, 0)$ is suitable to reproduce with high accuracy the slope matrix. Note that around $\omega = (-10, 10)$ the relative error in Figure 3.7 reaches one third of the exact value.

3.4.2 Systems with Wide Bandwidth

In systems with a large characteristic bandwidth, relative to v_0 and the Fermi level, problems may occur in the numerical determination of the slope matrix by the Taylor expansion (3.21). First, for v_0 lying far below the Fermi energy, *i.e.* for $\epsilon_F - v_0 \sim 1.0$ Ry or larger, the Taylor expansion (3.21) diverges for energies near and above ϵ_F. The second problem arises for bands, where $\epsilon_b \ll v_0$. Then the Taylor expansion breaks down for energies near the bottom of the valence band. Such a situation occurs, *e.g.*, in the case of early transition metals, some of the p metals, light actinides, oxides and nitrides, *etc.*, where the high-lying core states, the so called semi-core states are treated like the

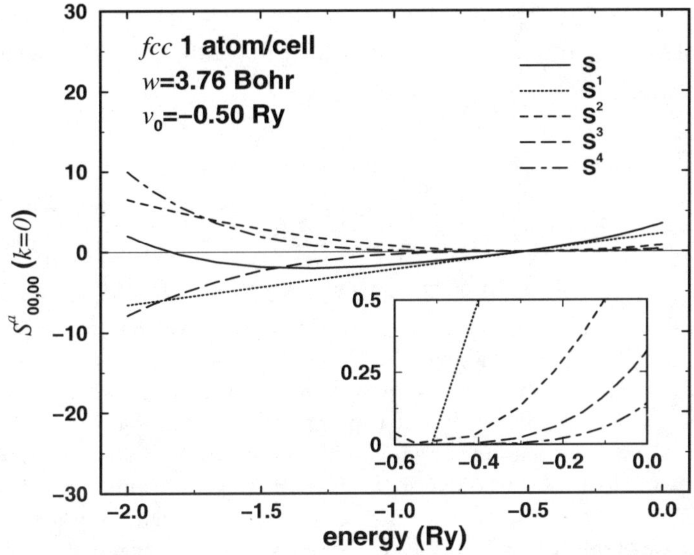

Fig. 3.8. The *ss* element of the *fcc* slope matrix and its first four energy derivatives calculated in the $\mathbf{k} = (0,0,0)$ point from the Brillouin zone. The matrix elements are plotted as functions of energy. The insert shows the behavior near the Fermi level at 0 Ry. S is the slope matrix calculated from the Taylor expansion (3.21), and S^i is the *i*th term from the Taylor expansion (see text). The Wigner–Seitz radius corresponds to Y, and v_0 is the constant potential.

valence states[3]. In these systems, the semi-core states are located with a few Rydbergs below the Fermi level. A typical energy window for these elements is approximately -2.5 Ry below v_0 and 1 Ry above v_0. Both the above problems become more pronounced in solids with a large w ($w^2 > 10$ Bohr^2).

As an example, we consider Y with atomic number $Z = 39$. Near the equilibrium volume, corresponding to Wigner–Seitz radius of 3.76 Bohr, the $4p^6$ semi-core states are located with ~ 1.8 Ry below the Fermi level. On the energy scale with origin pinned to the Fermi level, the energy window of interest is between $\epsilon_b \approx -2.0$ Ry and 0.2 Ry. To understand how the slope matrix behaves within this energy interval, we make use of the *fcc* slope matrix calculated at the Γ point from the reciprocal space and set up a fourth order expansion around $\omega_0 = 0$. The slope matrix in ω_0 and its first four derivatives were calculated in a similar way to that in the preceding section. The constant potential was fixed at -0.5 Ry, which is a typical v_0 for early transition metals (relative to the Fermi level). In the following, like in Sections 3.2 and 3.3, we concentrate only on the diagonal $L' = L = L_0$ element of the

[3] The energy levels of the core states lying close to the valence states might be altered significantly by the crystal potential. These states are usually refereed to as the semi-core states.

slope matrix. We use the notation S for the *ss* element calculated from the Taylor expansion, and S^i for the energy dependent terms from the Taylor series, *i.e.*

$$\begin{aligned} \mathrm{S} &= S^a_{RL_0RL_0}(\omega_0, \mathbf{k}_0) + \mathrm{S}^1 + \mathrm{S}^2 + \mathrm{S}^3 + \mathrm{S}^4, \\ \mathrm{S}^i &= \frac{1}{i!}\frac{d^i S^a_{RL_0RL_0}(\omega_0, \mathbf{k}_0)}{d\omega^i}(\omega - \omega_0)^i, \end{aligned} \quad (3.25)$$

where $S^a_{RL_0RL_0}(\omega_0, \mathbf{k}_0) \approx 0.0003$ (see Figure 3.5). Obviously, for a convergent series, S^i should decrease with the order of derivative. For the present case, S and S^i for i =1−4 are shown in Figure 3.8 as a function of energy. We observe that near v_0 the expansion converges rapidly. However, for energies well below v_0, the S^is increase when going from the first order to the fourth order derivative. This leads to the incorrect upturn of S with decreasing energy. Note also the slow convergence of the Taylor expansion near the Fermi level (insert), which is a consequence of to the relatively large volume.

The most straightforward way to improve the convergence of the expansion for a large energy window is to include a second Taylor expansion around a large negative ω. However, joining the two Taylor expansions usually creates problems in alloys, where the band gap between the valence and semi-core states contains other states. Alternatively, we may introduce additional screening spheres in the system. By this, we (a) improve on the localization of the screened spherical waves, and (b) reduce the interstitial region, and thus the characteristic length scale of the system.

Extra Screening Sites

In close-packed systems, the potential spheres can usually be chosen in such a way that the error coming from the potential sphere overlaps is small. In a simple fcc lattice, for instance, there is one hard sphere and one potential sphere per primitive cell. The optimal overlapping muffin-tin potential is centered on the lattice site and it has a radius close to the average atomic radius. This choice gives the minimal muffin-tin discontinuity $v_R(s_R) - v_0$. We denote by A the lattice sites belonging to such a parent lattice. The kink cancelation Equation (2.34) for this system has the following matrix form

$$K^a_{AA} g^a_{AA} \equiv a_A [S^a_{AA} - D^a_A] g^a_{AA} = I_{AA}, \quad (3.26)$$

where I_{AA} is the $N_A(l_{max}+1)^2 \times N_A(l_{max}+1)^2$ unitary matrix, N_A is the number of atomic sites per primitive cell (*i.e.* $N_A = 1$ for the original fcc lattice). For the sake of simplicity, in this section the L subscripts are omitted and the matrix multiplication is implied.

Next, we introduce N_E additional hard spheres in the system centered on the E interstitial positions. Then the average Wigner–Seitz radius becomes $w' = (1 + N_E)^{-1/3} w$. For instance, in the case of fcc Y, we put three hard spheres in the interstitial positions $(1/4, 1/4, 1/4)$, $(1/2, 1/2, 1/2)$ and

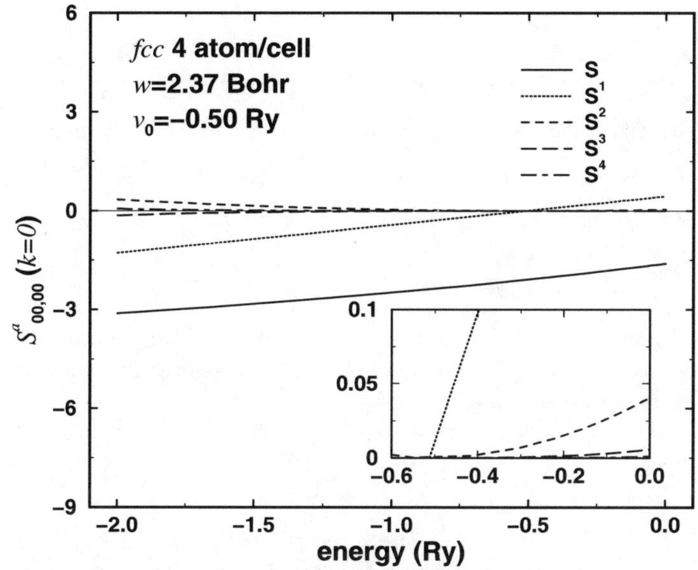

Fig. 3.9. The ss element of the slope matrix and its first four energy derivatives calculated at the $\mathbf{k} = (0,0,0)$ point from the Brillouin zone for an fcc lattice with basis. The four sites are located in the $(0,0,0)$, $(1/4,1/4,1/4)$, $(1/2,1/2,1/2)$ and $(3/4,3/4,3/4)$ positions from the unit cell. For notation see caption for Figure 3.8.

$(3/4,3/4,3/4)$ from the fcc lattice, so we arrive to the bcc packing with $w' = 2.37$ Bohr. For this system, we calculated the slope matrix and its derivatives at $\omega_0 = 0$ using real space clusters of 89 sites centered on the A and E sites. We used $l_{max} = 2$, $l^w_{max} = 8$ and the new hard sphere radii were $b_R = 0.70w'$. After the Bloch sum, for the on-site part of the slope matrix in the Γ point, we set up a fourth order Taylor expansion. The new S and S^i parameters, obtained according to Equation (3.25), are shown in Figure 3.9. A direct comparison between this figure and Figure 3.8 is not possible, because the hard sphere radii used to generate the two sets of data are different. Nevertheless, from Figure 3.9 we can see that the Taylor expansion for the new slope matrix converges well for any energy. The fourth order term is already negligible, even for energies near ϵ_b and for energies close to the Fermi level. This is a clear improvement compared to Figure 3.8.

When introducing the new screening spheres, we would like to keep the optimized overlapping potential in the original form. This can be done if we let the radius of the spherical potentials on the E sites vanish. For $s_E \to 0$, the logarithmic derivative on the E potential sphere and its energy derivative become

$$\mathcal{D}\{\phi_{El}(\epsilon, s_E)\} = \mathcal{D}\{j_l(\kappa s_E)\} \to l, \tag{3.27}$$

and

$$\dot{\mathcal{D}}\{\phi_{El}(\epsilon, s_E)\} = \dot{\mathcal{D}}\{j_l(\kappa s_E)\} \to 0. \tag{3.28}$$

The logarithmic derivative at the hard sphere radius b_E is calculated from Equation (2.27) as

$$D^b_{El}(\epsilon) = -\frac{f^b_{El}(\kappa^2, s_E)}{g^b_{Rl}(\kappa^2, s_E)} \frac{l - \mathcal{D}\{f^b_{El}(\kappa^2, s_E)\}}{l - \mathcal{D}\{g^b_{El}(\kappa^2, s_E)\}} = \frac{t^1_{El}(\kappa^2)}{t^3_{El}(\kappa^2)}, \tag{3.29}$$

where we have used the properties of the Bessel and Neumann functions (Equations (B.22) and (B.23)) near the origin. The kink cancelation equation for the new system has the form

$$\begin{pmatrix} K^b_{AA} & K^b_{AE} \\ K^b_{EA} & K^b_{EE} \end{pmatrix} \begin{pmatrix} g^b_{AA} & g^b_{AE} \\ g^b_{EA} & g^b_{EE} \end{pmatrix} = \begin{pmatrix} I_{AA} & 0 \\ 0 & I_{EE} \end{pmatrix} \tag{3.30}$$

with $K^b = b(S^b - D^b)$. Here S^b is the slope matrix for the $A+E$ lattice and D^b is the proper logarithmic derivative for the A sites and reduces to (3.29) for the E sites. I_{EE} is the $N_E(l_{max}+1)^2 \times N_E(l_{max}+1)^2$ unitary matrix. When sufficient numbers of E sites are introduced, the new slope matrix converges well even for a very large bandwidth. The solution of Equation (3.30) gives the charge density around the A sites. Note that the radial Schrödinger equation (2.20) and the single-electron potential (2.70) have to be solved only for the A sites, but the kink cancelation equation contains both the A and E sites. In practice, this limits N_E to a relatively small number.

Before closing this section, we show that Equation (3.30) is equivalent to the original problem (3.26). For this, we explicitly write the elements of the new kink matrix

$$\begin{aligned} K^b_{AA} &= b_A[S^b_{AA} - D^b_A] \ , \ K^b_{AE} = b_A S^b_{AE} \\ K^b_{EA} &= b_E S^b_{EA} \ , \ K^b_{EE} = b_E[S^b_{EE} - D^b_E]. \end{aligned} \tag{3.31}$$

Since D^b_E does not depend on the potential, the kink cancelation may be solved for g^b_{AA} using the equation

$$K^b_{AA} g^b_{AA} + K^b_{AE} g^b_{EA} = I_{AA} \tag{3.32}$$

obtained from Equation (3.30). In addition to the AA block, this expression contains g^b_{EA} that couples the original atomic sites A with the additional screening sites E. From Equation (3.30) we have

$$K^b_{EA} g^b_{AA} + K^b_{EE} g^b_{EA} = 0, \tag{3.33}$$

so

$$g^b_{EA} = -(K^b_{EE})^{-1} K^b_{EA} g^b_{AA} = -[S^b_{EE} - D^b_E]^{-1} S^b_{EA} g^b_{AA}. \tag{3.34}$$

After eliminating g^b_{EA} from Equations (3.32) and (3.34), we obtain

$$[K^b_{AA} - K^b_{AE}(K^b_{EE})^{-1} K^b_{EA}] g^b_{AA} = I_{AA}. \tag{3.35}$$

Using Equation (3.31), the final KKR problem becomes

$$b_A[\tilde{S}^b_{AA} - D^b_A] g^b_{AA} = I_{AA}, \tag{3.36}$$

where

$$\tilde{S}^b_{AA} = S^b_{AA} - S^b_{AE}[S^b_{EE} - D^b_E]^{-1} S^b_{EA}. \tag{3.37}$$

Equation (3.36) is formally equivalent to the original KKR problem (3.26) written for the lattice without extra hard spheres. The slope matrix for the reduced system is obtained from the slope matrix of the $A+E$ system by the potential independent transformation (3.37), where D^b_E is given in Equation (3.29). Note that Equations (3.26) and (3.36) are given in different representations. Writing the Dyson equation (3.11) for the combined $A+E$ system and for hard sphere radii b_R, we find

$$\begin{aligned} &- \begin{pmatrix} S^0_{AA} + t^1_A/t^3_A & S^0_{AE} \\ S^0_{EA} & S^0_{EE} + t^1_E/t^3_E \end{pmatrix} \begin{pmatrix} t^3_A S^b_{AA} - t^1_A t^3_A & t^3_A S^b_{AE} \\ t^3_E S^b_{EA} & t^3_E S^b_{EE} - t^1_E t^3_E \end{pmatrix} \\ &= \begin{pmatrix} I_{AA} & 0 \\ 0 & I_{EE} \end{pmatrix}. \end{aligned} \tag{3.38}$$

Eliminating the off-diagonal blocks from the first matrix, and using the fact that for the original lattice with hard spheres b_R we have

$$-\left(S^0_{AA} + t^1_A/t^3_A\right)\left(t^3_A \tilde{S}^b_{AA} - t^1_A t^3_A\right) = I_{AA}, \tag{3.39}$$

we arrive at Equation (3.37). By this, we have demonstrated that Equation (3.30) written for the system formed by the $A+E$ sites is equivalent to the original Equation (3.26). However, in order to avoid the convergence problems near the energies far from v_0, one should always use (3.30) rather than (3.26). This is because, the screening is improved due to the extra hard spheres and as a consequence the energy derivatives of S^b are computed with higher numerical accuracy than those for the parent lattice.

3.4.3 Two-center Expansion

In the previous section, we have shown that accurate energy derivatives can improve the Taylor expansion. Therefore, we could remove the bad behavior at energies around the bottom of the band by setting up the higher order energy derivatives in a more accurate way. Increasing the size of the real space cluster is not a feasible solution, because the number of surface resonances increase with the surface area. Another solution would be to generate the higher order derivatives using data calculated for a different energy. This can be formulated, *e.g.*, as a two-center expansion. We consider two distinct energy points ω_0 and ω_1 where the value and derivatives of S^a are known. We expand $S^a(\omega)$ in such a way that the expansion should reproduce exactly the first n derivatives of S^a in ω_0 and the first m derivatives of S^a in ω_1. Mathematically, this can be formulated as

$$\begin{aligned} S^a(\omega) &\approx S^{n,m}(\omega;\omega_0,\omega_1) \\ &= a_0 + \frac{1}{1!}a_1(\omega-\omega_0) + \frac{1}{2!}a_2(\omega-\omega_0)^2 + ... + \frac{1}{n!}a_n(\omega-\omega_0)^n \\ &+ \frac{1}{(n+1)!}a_{n+1}(\omega-\omega_0)^{n+1} + \frac{1}{(n+2)!}a_{n+2}(\omega-\omega_0)^{n+2} + ... \\ &+ \frac{1}{(n+m)!}a_{n+m+1}(\omega-\omega_0)^{n+m+1}, \end{aligned} \tag{3.40}$$

where, for simplicity, we have dropped the RL subscripts. Obviously, for the first $(n+1)$ coefficients we have

$$a_i = \frac{d^i S^a(\omega_0)}{d\omega^i} \quad \text{for} \quad i = 0, 1, 2, ..., n. \tag{3.41}$$

The last $(m+1)$ coefficients are obtained from the conditions

$$\begin{aligned} S^{n,m}(\omega_1;\omega_0,\omega_1) &= S^a(\omega_1), \\ \frac{dS^{n,m}(\omega;\omega_0,\omega_1)}{d\omega}|_{\omega=\omega_1} &= \frac{dS^a(\omega)}{d\omega}|_{\omega=\omega_1}, \\ &\cdots \\ \frac{d^m S^{n,m}(\omega;\omega_0,\omega_1)}{d\omega^m}|_{\omega=\omega_1} &= \frac{d^m S^a(\omega)}{d\omega^m}|_{\omega=\omega_1}. \end{aligned} \tag{3.42}$$

These conditions lead to a system of linear equations for $a_{n+1}, a_{n+2}, ...$. Solving these equations, we obtain the $(n+m+1)$th order expansion for S^a.

The two-center expansion we demonstrate in the case of Y in *fcc* lattice. The two slope matrices and the first four energy derivatives were calculated for $\omega_0 = 0$ and $\omega_1 = -30$ using $l_{max} = 2$, $l_{max}^w = 8$, a real space cluster of 79 sites and $a_R = 0.70w$. The Wigner–Seitz radius and constant potential were

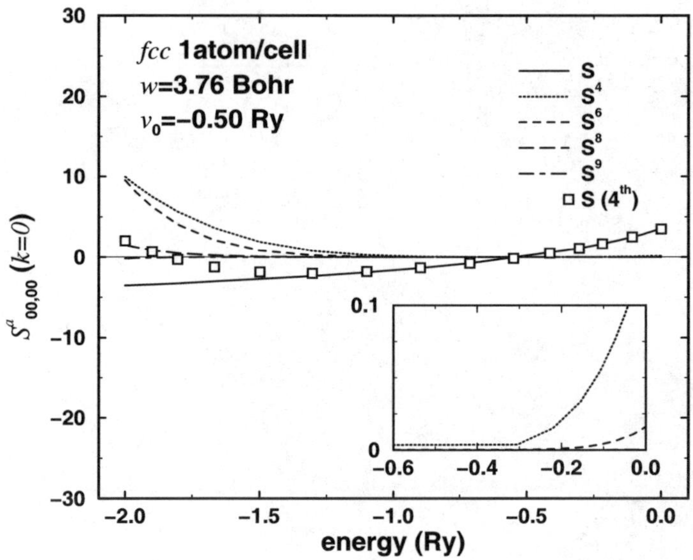

Fig. 3.10. The *ss* element of the *fcc* slope matrix and its four energy derivatives calculated in the $\mathbf{k} = (0, 0, 0)$ point from the Brillouin zone. For notation see caption for Figure 3.8.

set to 3.76 Bohr and $\epsilon_F - v_0 = 0.5$, respectively. In Figure 3.10, we show S, S^4, S^6, S^8 and S^9, calculated from a_4, a_6, a_8 and a_9 according to Equation (3.25). For comparison, we also included in the figure the slope matrix obtained from the fourth order Taylor expansion (Figure 3.8). We can immediately see that the 9th order expansion is highly accurate for any energy. The 8th and 9th order terms are already negligible over the entire energy region. The large diverging derivative terms obtained from a fourth order Taylor expansion (see also Figure 3.8) are canceled by the 5th, 6th and 7th order terms. In fact, it has turned out that using a two-center expansion with $n \approx m \approx 6$, for an average $w^2 \approx 10$ Bohr^2 one can accurately map an energy window as large as 4-6 Ry below and ~ 0.5 Ry above the Fermi level.

4 Full Charge Density Technique

According to the Hohenberg–Kohn variational principle [1], the total energy functional is stationary for small density variations around the equilibrium density. Therefore, a reasonably accurate trial density is suitable to determine the total energy of the system within an error which is second order in the difference between the trial and equilibrium charge densities. This recognition has led to the elaboration of the *Full Charge Density* (FCD) technique [44, 45, 48, 49] as an alternative to the full-potential methods. The FCD technique is designed to maintain high efficiency but at the same time to give total energies with an accuracy similar to that of the full-potential methods. It assumes the knowledge of just the spherically symmetric part of the potential but at the same time makes use of the full non-spherically symmetric charge density. In recent years it turned out that results obtained from such a technique compare very well to those of full potential methods. Today many research groups adopt this technique in combination with a muffin-tin type of method rather than the formally exact but very demanding full-potential approach [42, 43, 44, 45, 77, 92, 93].

The principal idea behind the FCD technique is to use the total charge density to compute the total energy functional given by Equations (1.9) and (1.12). The total density can be taken from a self-consistent calculation employing certain approximations. In the present case we use the EMTO total charge density (2.47) written in the one-center form (2.48). In order to be able to compute the energy components from Equation (1.9) we need to establish a technique to calculate the space integrals over the Wigner–Seitz cells. For this we adopt the *shape function technique* [44]. The interaction energy between remote Wigner–Seitz cells is taken into account through the Madelung term. A particularly delicate contribution to this energy arises from Wigner–Seitz cells with overlapping bounding spheres. This energy is calculated by the so called *displaced cell technique* [94, 95].

The shape function technique will be introduced in Section 4.1. Here we shall present an algorithm which is suitable for determining the shape function for an arbitrary crystal structure. Using the shape function formalism, in Section 4.2 we shall give the expression for the FCD total energy. The displaced cell technique will be presented in Section 4.2.5. At the end of this chapter, the convergence properties of the energy components will be discussed.

4.1 Shape Function Technique

There is a large number of numerical techniques used to carry out the 3-dimensional (3D) integrations over the Wigner–Seitz cells [27, 48, 96, 97, 98]. Here we employ the shape function or truncation function technique originally introduced by Andersen and Woolley [96]. This approach has also been implemented in different full-potential Korringa–Kohn–Rostoker multiple scattering methods [22, 93].

By means of the shape function any integral over the cell can be transformed into an integral over the sphere which circumscribes the cell. The shape function is a 3D step function defined as 1 inside the Wigner–Seitz cell (Ω_R) and zero otherwise, *i.e.*

$$\sigma_R(\mathbf{r}_R) \equiv \begin{cases} 1 & \text{for } \mathbf{r}_R \in \Omega_R \\ 0 & \text{otherwise} \end{cases} . \tag{4.1}$$

At each point on the radial mesh r_R the shape function is expanded in terms of real harmonics

$$\sigma_R(\mathbf{r}_R) = \sum_L \sigma_{RL}(r_R) Y_L(\hat{r}_R). \tag{4.2}$$

The functions

$$\sigma_{RL}(r_R) \equiv \int \sigma_R(\mathbf{r}_R) Y_L(\hat{r}_R) d\hat{r}_R \tag{4.3}$$

are the partial components of the shape function. They constitute the needed description of the Wigner–Seitz cell and contain all dependence of the shape function on the cell shape.

Once the partial components have been evaluated for a given cell, any integral over the cell can be transformed into an integral over the sphere which circumscribes the cell. We denote by s_R^c the radius of the smallest circumscribed or bounding sphere centered on lattice site R. Then the integral over the Wigner–Seitz cell Ω_R of an arbitrary functional of the electron density $K([n]; \mathbf{r}_R)$ can be expressed as

$$\int_{\Omega_R} n_R(\mathbf{r}_R)\, K([n]; \mathbf{r}_R)\, d\mathbf{r}_R = \int_{s_R^c} \sigma_R(\mathbf{r}_R)\, n_R(\mathbf{r}_R)\, K([n]; \mathbf{r}_R)\, d\mathbf{r}_R. \tag{4.4}$$

We expand the functions $\sigma_R(\mathbf{r}_R)\, n_R(\mathbf{r}_R)$ in terms of real harmonics, *viz.*

$$\sigma_R(\mathbf{r}_R)\, n_R(\mathbf{r}_R) \;=\; \sum_L \tilde{n}_{RL}(r_R) Y_L(\hat{r}_R). \tag{4.5}$$

The radial function $\tilde{n}_{RL}(r_R)$ represents the $Y_L(\hat{r}_R)$ projection of the charge density on a spherical surface that lies inside the Wigner–Seitz cell. In terms of the partial components of the shape function and charge density, the latter can be expressed as

$$\tilde{n}_{RL}(r_R) \;=\; \sum_{L',L''} C_{LL'L''} n_{RL'}(r_R) \sigma_{RL''}(r_R), \tag{4.6}$$

where $C_{LL'L''}$ are the real Gaunt coefficients. Now, if $K([n];\mathbf{r}_R)$ is also expanded in terms of the real harmonics, the integral over the Wigner–Seitz cell has a particularly simple expression

$$\int_{\Omega_R} n_R(\mathbf{r}_R)\, K([n];\mathbf{r}_R)\, d\mathbf{r}_R \;=\; \sum_L \int_0^{s_R^c} \tilde{n}_{RL}(r_R) K_L(r_R) r_R^2 dr_R, \tag{4.7}$$

where

$$K_L(r_R) \;\equiv\; \int K([n];\mathbf{r}_R) Y_L(\hat{r}_R) d\hat{r}_R \tag{4.8}$$

is the $Y_L(\hat{r}_R)$ projection of $K([n];\mathbf{r}_R)$ on the spherical surface with radius r_R.

4.1.1 Numerical Calculation of the Shape Function

The partial components of the shape function are calculated from Equation (4.3), or equivalently from

$$\sigma_{RL}(r_R) \;=\; \int_{\mathcal{S}_R(r_R)} Y_L(\hat{r}_R) d\hat{r}_R \tag{4.9}$$

where $\mathcal{S}_R(r_R)$ represents that part of the spherical surface of radius r_R which lies inside the Wigner–Seitz cell.

In Table 4.1, we list the inscribed, circumscribed and atomic sphere radii for the body centered cubic (*bcc*), face centered cubic (*fcc*) and simple cubic (*sc*) lattices. Obviously, inside the inscribed sphere the surface integration in (4.9) is performed on the complete solid angle and thus for the partial components we get

$$\sigma_{RL}(r_R) \;=\; \sqrt{4\pi}\, \delta_{l0} \;\text{ for }\; r_R \le s_R^i. \tag{4.10}$$

For radii larger than the circumscribed sphere $\mathcal{S}_R(r_R)$ disappears and we have

$$\sigma_{RL}(r_R) \;=\; 0 \;\text{ for }\; r_R \ge s_R^c. \tag{4.11}$$

Table 4.1. The inscribed (s^i), circumscribed (s^c) and atomic radii (w) for *bcc*, *fcc* and *sc* primitive unit cells. The radii are expressed in units of lattice constant a.

	bcc	*fcc*	*sc*
s^i	0.433013	0.353553	0.5
s^c	0.559017	0.5	0.866025
w	0.492373	0.390796	0.620350

To calculate the partial shape functions for intermediate radii, we divide the Wigner–Seitz cell into tetrahedra. This is described in more detail below. Then the surface integral from Equation (4.9) may also be divided into solid angles corresponding to different tetrahedra and the total shape function is obtained by summing up the individual contributions. Since many of these tetrahedra will be equivalent the surface integral has to be carried out only for a few non-equivalent tetrahedra.

The Wigner–Seitz cell is a convex polyhedron bounded by N_f planes drawn perpendicularly at the midpoints of the vectors connecting neighboring lattice sites. This polyhedron can be divided into N_f pyramids, each of them having as base one of the N_f facets and apex set on the actual lattice site. The base of the pyramid p (p runs from 1 to N_f) is a polygon with $N_e(p)$ sides. In the case of fcc lattice, we have $N_f = 12$ and all the polygons are quadrilaterals (*i.e.*, $N_e(p) = 4$). For *bcc* lattice, eight out of the $N_f = 14$ polygons are hexagons and six are quadrilaterals. For *sc* lattice, we have $N_f = 6$ and all the polygons are squares.

The line going through the apex and perpendicular to the base of the pyramid together with the $2 \times N_e(p)$ edges divide the pyramid into $N_e(p)$ tetrahedra. Each of these tetrahedra has the height of the pyramid and two of the three lateral facets are perpendicular to the base. Totally there are $N_t = \sum_{p=1}^{N_f} N_e(p)$ such tetrahedra, but usually only a few of them are non-equivalent. We denote by N_n the number of non-equivalent tetrahedra and by $N_e(t)$ the number of tetrahedra of type t, *i.e.* $\sum_{t=1}^{N_n} N_e(t) = N_t$. For close-packed crystals, usually we have $N_n \ll N_t$. For instance, for *bcc*, fcc and *sc* lattices there are $2, 2$ and 1 non-equivalent tetrahedra, respectively.

According to the above division, the total shape function is obtained as

$$\sigma_{RL}(r_R) = \sum_t^{N_n} \sum_i^{N_e(t)} \left\{ \sum_{m'} D^l_{m\,m'}(\alpha_i, \beta_i, \gamma_i) \sigma^t_{Rlm'}(r_R) \right\}, \tag{4.12}$$

where $D^l_{m\,m'}$ are the matrix elements of finite rotations defined in Appendix B. For each tetrahedron i we choose a local coordinate system. The origin of this system is set on the apex (*i.e.* the actual lattice site), the x axis is oriented along the edge normal to the base and the xy plane is chosen to be one of the facets which is perpendicular to the base. In Equation (4.12), α_i, β_i

and γ_i are the Euler angles of this local coordinate system for tetrahedron i relative to the original (common) coordinate system.

The partial shape functions for the tetrahedron t are calculated within the the local coordinate system, *viz.*

$$\sigma^t_{Rlm}(r_R) = \int_{\varphi^t_{min}}^{\varphi^t_{max}} \left\{ \int_{\theta^t_{min}(\varphi)}^{\theta^t_{max}(\varphi)} Y_{lm}(\theta,\varphi)\sin\theta d\theta \right\} d\varphi. \tag{4.13}$$

The integration intervals for φ and θ are specified by φ^t_{min} and φ^t_{max} and by $\theta^t_{min}(\varphi)$ and $\theta^t_{max}(\varphi)$, respectively, which can easily be determined from the geometry of the tetrahedron. The integration over θ is performed analytically using the procedure proposed by Stefanou *et al.* [98]. From expressions (B.10) and (B.11) the θ-integration becomes

$$\int Y_{lm}(\theta,\varphi)\sin\theta d\theta = \frac{1}{\sqrt{4\pi}} y_m(\varphi) \sum_{k\geq(l+|m|)/2}^{l} H^{|m|}_{lk} F^{2k-l-|m|}_{|m|+1}(\theta) \tag{4.14}$$

where $y_m(\varphi)$ and $H^{|m|}_{lk}$ are defined in Appendix B. The new integrals of θ,

$$F^q_p(\theta) \equiv \int \sin^p\theta\cos^q\theta d\theta, \tag{4.15}$$

can be computed from the recurrence relations

$$\begin{aligned}(p+q)F^q_p(\theta) &= (p-1)F^q_{p-2}(\theta) - \sin^{p-1}\theta\cos^{q+1}\theta \\ &= (q-1)F^{q-2}_p(\theta) + \sin^{p+1}\theta\cos^{q-1}\theta\end{aligned} \tag{4.16}$$

with the initial values

$$\begin{aligned}F^0_0(\theta) &= \theta,\ \ F^0_1(\theta) = -\cos\theta, \\ F^1_0(\theta) &= \sin\theta \text{ and } F^1_1(\theta) = -\frac{1}{2}\cos^2\theta.\end{aligned} \tag{4.17}$$

From Equation (4.14) for the partial shape functions (4.13) we obtain

$$\begin{aligned}\sigma^t_{Rlm}(r_R) = \frac{1}{\sqrt{4\pi}} \sum_{k\geq(l+|m|)/2}^{l} H^{|m|}_{lk} \int_{\varphi^t_{min}}^{\varphi^t_{max}} y_m(\varphi) \\ \times \left\{ F^{2k-l-|m|}_{|m|+1}\left(\theta^t_{max}(\varphi)\right) - F^{2k-l-|m|}_{|m|+1}\left(\theta^t_{min}(\varphi)\right)\right\} d\varphi.\end{aligned} \tag{4.18}$$

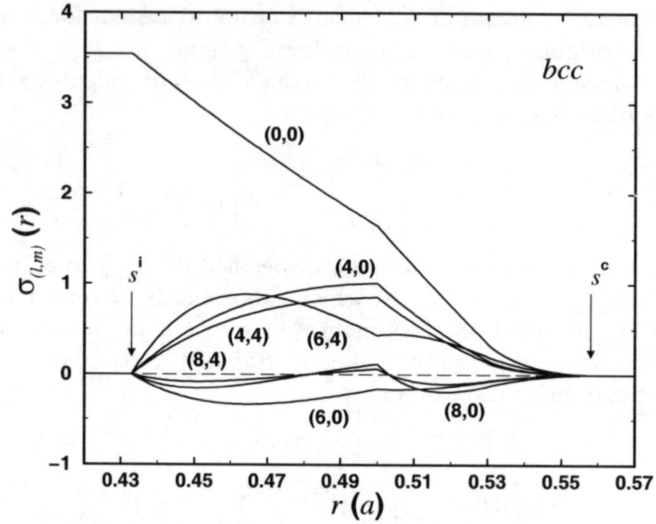

Fig. 4.1. The partial components of the *bcc* shape function for $(l,m) = (0,0),(4,0),(4,4),(6,0),(6,4),(8,0)$ and $(8,4)$ as functions of the radius r (in units of lattice constant).

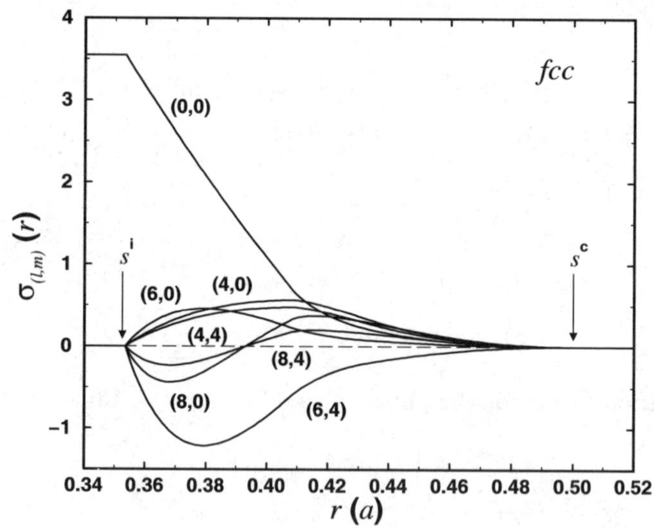

Fig. 4.2. The partial components of the fcc shape function for $(l,m) = (0,0),(4,0),(4,4),(6,0),(6,4),(8,0)$ and $(8,4)$ as functions of the radius r (in units of lattice constant).

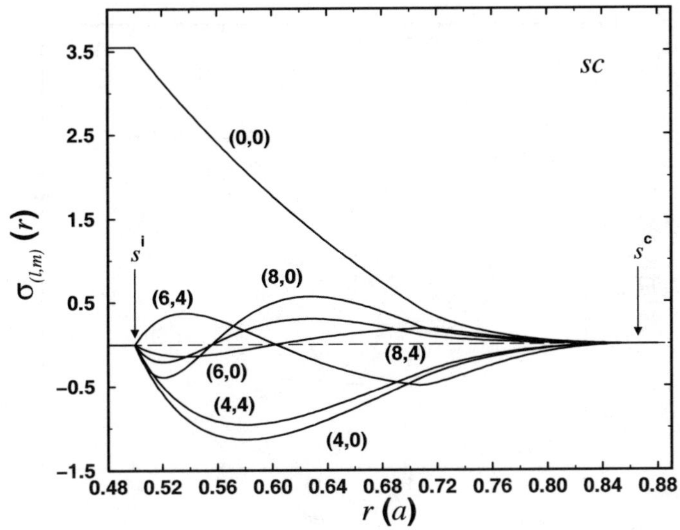

Fig. 4.3. The partial components of the *sc* shape function for $(l,m) = (0,0),(4,0),(4,4),(6,0),(6,4),(8,0)$ and $(8,4)$ as functions of the radius r (in units of lattice constant).

The integration over φ is carried out using a numerical integration method. Equations (4.16)−(4.18) together with Equation (4.12) are suitable to compute the shape function of an arbitrary crystal structure with a high accuracy and efficiency.

In Figures 4.1−4.3 we show some of the partial components of the shape functions for the *bcc*, *fcc* and *sc* structures, respectively. The shape functionals are shown as functions of the radial distance r expressed in units of the cubic lattice constant a. Apart from the spherical component $\sigma_{(0,0)}(r)$, all the other terms are zero inside the inscribed sphere.

The partial shape functions have kinks at the points where the sphere of radius r passes through a new facet of the polyhedron. This can be seen, for instance, in Figure 4.1, where for $r = 0.5a$ the sphere touches the square facets of the *bcc* polyhedron. Discontinuities in the higher order derivatives appear at the points where the sphere passes through a new edge and vertex of the polyhedron. This happens for $r = 3\sqrt{2}/8$, $r = \sqrt{6}/6$, $\sqrt{3}/4$ and $r = \sqrt{2}/2$ in *bcc*, *fcc* and *sc* lattices, respectively. In order to ensure the required accuracy in the numerical integration over the radial mesh, these points should be taken as mesh points.

4.1.2 The l-convergence of the Shape Function

In Equation (4.2), the number of partial shape functions is infinite. In practice, however, the l-summation has to be truncated at a reasonably low l^s_{max}. Usu-

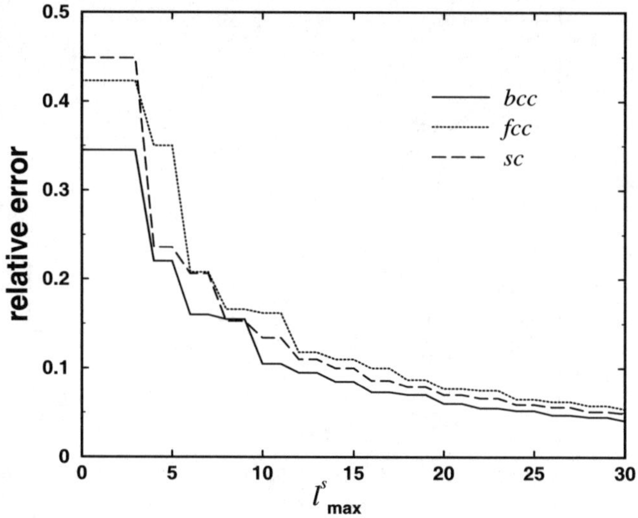

Fig. 4.4. The relative errors between the volumes of the *bcc*, *fcc* and *sc* Wigner–Seitz cells and the corresponding inscribed spheres as functions of the maximum l used in expansion (4.2).

ally, l^s_{max} is chosen to be less than $\sim 30-40$ but significantly larger than l^h_{max}, which is the l-truncation of the one-center expansion of the charge density, *i.e.* the largest l in expansion (2.48). Then, the l-truncation in Equation (4.5) will be $l^s_{max} + l^h_{max} \approx l^s_{max}$, which also gives the maximum number of terms included in integrals such as the one from the right hand side of Equation (4.7).

The shape function calculated from Equation (4.2) oscillates strongly as a function of l^s_{max} and its convergence towards the exact step function (4.1) is rather slow. To illustrate this, we consider the volume between the Wigner–Seitz cell and the inscribed sphere

$$\Delta\Omega^i \equiv 4\pi w^3/3 - 4\pi(s^i)^3/3. \tag{4.19}$$

This quantity may also be calculated as

$$\int_{s^c} \sigma(\mathbf{r})\, \sigma(\mathbf{r})\, d\mathbf{r} - \int_{s^i} \sigma(\mathbf{r})\, \sigma(\mathbf{r})\, d\mathbf{r} = \sum_L \int_{s^i}^{s^c} \sigma_L^2(r)\, r^2\, dr \equiv \sum_l^{l^s_{max}} d_l, \tag{4.20}$$

where for $\sigma(\mathbf{r})$ we used expansion (4.2) truncated at l^s_{max}. Obviously, when $l^s_{max} \to \infty$ we have $\sum_l^{l^s_{max}} d_l \to \Delta\Omega^i$. Figure 4.4 shows the relative error

$$\delta\Delta\Omega^i(l^s_{max}) = (\Delta\Omega^i - \sum_l^{l^s_{max}} d_l)/\Delta\Omega^i \tag{4.21}$$

as a function of l^s_{max}. As can be seen, for all three cubic structures the relative errors are still very large ($\sim 5\%$) for $l^s_{max} = 30$. On the other hand, the partial components $\sigma_L(r)$ exhibit several oscillations within the interval $s^i \leq r \leq s^c$ and the number of oscillations increases with the orbital quantum number (see Figures 4.1–4.3). Therefore, the quantities derived from the shape function by integrations like in Equation (4.4) or (4.7) are expected to show a faster l^s_{max} convergence compared to $\delta\Delta\Omega^i(l^s_{max})$. This will be demonstrated in Section 4.2.6.

4.2 The FCD Total Energy

Our aim is to evaluate the total energy functional from Equations (1.9) and (1.12) using the total charge density obtained from a self-consistent EMTO calculation. We assume that the total density is decomposed into cell contributions according to Equation (2.47) and within each unit cell it has the one-center form (2.48). First, in E_{tot} we separate the kinetic energy, T_s, the exchange-correlation energy, E_{xc}, and the electrostatic energy, E_c. The latter is given by

$$\begin{aligned} E_c[n] &= \int\int \frac{n(\mathbf{r}')n(\mathbf{r})}{|\mathbf{r}-\mathbf{r}'|} d\mathbf{r}' d\mathbf{r} \\ &+ \int \left(-\sum_R \frac{2Z_R}{|\mathbf{r}-\mathbf{R}|} \right) n(\mathbf{r}) d\mathbf{r} + \sum_{RR'}{}' \frac{Z_R Z_{R'}}{|\mathbf{R}-\mathbf{R}'|}, \end{aligned} \tag{4.22}$$

where the prime indicates that only terms with $R \neq R'$ are included in the sum. Next, E_c is split into the intra-cell F_{intra} and inter-cell F_{inter} contributions. The prior is due to the charges inside a Wigner–Seitz cell and the latter is the interaction between different cells, which is commonly referred to as the Madelung energy. According to this division, the total energy becomes

$$E_{tot} = T_s[n] + \sum_R (F_{intraR}[n_R] + E_{xcR}[n_R]) + F_{inter}[n], \tag{4.23}$$

where we point out that the intra-cell and exchange-correlation energies depend only on the charge density within the actual cell[1], whereas F_{inter} depends on the charge distributions around different cells and T_s is a nonlocal functional of the density. In the following, we describe how the different energy contribution to E_{tot} are calculated within the FCD scheme.

[1] In fact, within a gradient level approximation, the exchange-correlation energy also depends on the electron density slightly beyond the cell boundary needed to compute the density gradient.

4.2.1 Kinetic Energy

Since the kinetic energy is an unknown functional of the density, its direct evaluation from $n(\mathbf{r})$ is not feasible. On the other hand, using the single-electron Kohn–Sham equations (1.3), $T_s[n]$ can be derived from the self-consistent single-electron energies and Kohn–Sham potential. Within the EMTO formalism from Equation (1.10) we obtain

$$T_s[n] = \frac{1}{2\pi i} \oint_{\epsilon_F} z\, G(z)\, dz - \sum_R \int_{\Omega_R} v_{mt}(\mathbf{r}_R) n_R(\mathbf{r}_R) d\mathbf{r}_R. \tag{4.24}$$

The first term from the right hand side is the sum of the single-electron energies and $G(z)$ is given in Equation (2.46). In the second term, $v_{mt}(\mathbf{r}_R)$ is the optimized overlapping muffin-tin potential. Using Equation (2.1), this term can be recast into the following simple form

$$-\sum_R \sqrt{4\pi} \int_0^{s_R} [v_R(r_R) - v_0] n_{RL_0}(r_R) r_R^2 dr_R - v_0 N_e, \tag{4.25}$$

where N_e is the total number of electrons from the unit cell and $L_0 = (0,0)$. This expression is not fully consistent with the EMTO overlap matrix in Equation (2.43) and the number of states in Equation (2.45), where the terms coming from the overlap region have been neglected. A more appropriate kinetic energy is obtained if the second term in Equation (4.24) is calculated directly within the unit cell. Denoting by $v_{mt}(r_R)$ the muffin-tin potential inside the Wigner–Seitz cell at R without taking into account the potential wells centered on the neighboring lattice sites, *i.e.*

$$v_{mt}(r_R) = \begin{cases} v_R(r_R) & \text{if } r_R \le s_R \\ v_0 & \text{if } r_R > s_R \end{cases}, \tag{4.26}$$

for the second term from the right hand side of Equation (4.24) we obtain

$$-\sum_R \sqrt{4\pi} \int_0^{s_R^c} v_{mt}(r_R) \tilde{n}_{RL_0}(r_R) r_R^2 dr_R, \tag{4.27}$$

where $\tilde{n}_{RL}(r_R)$ is defined in Equation (4.6). Note that in a numerical integration, the radial integral from (4.27) should be carried out separately for $0 \le r_R \le s_R$ and $s_R \le r_R \le s_R^c$.

4.2.2 Exchange-correlation Energy

The exchange-correlation energy belonging to the cell at R is written as the integral over the Wigner–Seitz cell of the exchange-correlation energy per electron, $\varepsilon_{xc}([n];\mathbf{r}) = \varepsilon_x([n];\mathbf{r}) + \varepsilon_c([n];\mathbf{r})$, multiplied with the electron density, *i.e.*

$$E_{xcR}[n_R] = \int_{\Omega_R} n_R(\mathbf{r}_R)\varepsilon_{xc}([n_R];\mathbf{r}_R)d\mathbf{r}_R. \tag{4.28}$$

A few commonly used approximations for the exchange and correlation energy densities are given in Appendix A. For charge densities which deviate weakly from spherical symmetry, the exchange-correlation energy density may be represented by Taylor series around the spherically symmetric charge density [45], and, therefore, the $3D$ integral may be reduced to a radial integral. However, for strongly anisotropic electron densities, like in the case of surfaces or open structures, the Taylor expansion is not convergent. In this case the exchange-correlation energy is evaluated by a direct $3D$ integration over the circumscribed sphere, *i.e.*

$$\begin{aligned} E_{xc}[n_R] &= \int_0^{2\pi}\int_0^{\pi}\int_0^{s_R^c} n_R(\mathbf{r}_R)\varepsilon_{xc}([n_R];\mathbf{r}_R) \\ &\times \sum_L^{l_{max}^s} \sigma_{RL}(r_R)Y_L(\hat{r}_R)\, r_R^2\, dr_R\, \sin\theta d\theta d\phi. \end{aligned} \tag{4.29}$$

The convergence properties of this expression will be discussed in Section 4.2.6.

4.2.3 Intra-cell Electrostatic Energy

The intra-cell energy belonging to the cell at R,

$$\begin{aligned} F_{intraR}[n_R] &= \int_{\Omega_R}\int_{\Omega_R} \frac{n_R(\mathbf{r}'_R)n_R(\mathbf{r}_R)}{|\mathbf{r}_R - \mathbf{r}'_R|}\, d\mathbf{r}'_R d\mathbf{r}_R \\ &- \int_{\Omega_R} \frac{2Z_R}{r_R}\, n_R(\mathbf{r}_R)d\mathbf{r}_R, \end{aligned} \tag{4.30}$$

may easily be evaluated by introducing the shape function and using expansion (2.59). After simple mathematics, we arrive at

$$\begin{aligned} F_{intraR}[n_R] &= \frac{\sqrt{4\pi}}{w}\sum_L \int_0^{s_R^c} \tilde{n}_{RL}(r_R)\left[\left(\frac{r_R}{w}\right)^l P_{RL}(r_R)\right. \\ &+ \left.\left(\frac{r_R}{w}\right)^{-l-1} Q_{RL}(r_R) - 2Z_R\frac{w}{r_R}\delta_{L,L_0}\right] r_R^2 dr_R, \end{aligned} \tag{4.31}$$

where $\tilde{n}_{RL}(r_R)$ is defined in Equation (4.6) and the two radial functions are

$$P_{RL}(r_R) \equiv \frac{\sqrt{4\pi}}{2l+1}\int_{r_R}^{s_R^c} \tilde{n}_{RL}(r'_R)\left(\frac{r'_R}{w}\right)^{-l-1}(r'_R)^2 dr'_R, \quad (4.32)$$

and

$$Q_{RL}(r_R) \equiv \frac{\sqrt{4\pi}}{2l+1}\int_0^{r_R} \tilde{n}_{RL}(r'_R)\left(\frac{r'_R}{w}\right)^{l}(r'_R)^2 dr'_R. \quad (4.33)$$

4.2.4 Inter-cell Electrostatic Energy

The inter-cell energy includes the electrostatic interactions between different cells, *i.e.*

$$F_{inter}[n] = \sum_{RR'}{}' \left\{ \int_{\Omega_R}\int_{\Omega_{R'}} \frac{n_{R'}(\mathbf{r}_{R'})n_R(\mathbf{r}_R)}{|\mathbf{r}_R - \mathbf{r}_{R'} + \mathbf{R} - \mathbf{R}'|} d\mathbf{r}_{R'}d\mathbf{r}_R - \int_{\Omega_R} \frac{2Z_{R'}n(\mathbf{r}_R)}{|\mathbf{r}_R + \mathbf{R} - \mathbf{R}'|} dr_R + \frac{Z_R Z_{R'}}{|\mathbf{R}-\mathbf{R}'|} \right\}, \quad (4.34)$$

where the prime on the summation represents the restriction $R \neq R'$. For cells having non-overlapping bounding spheres, Equation (4.34) is calculated using expansions (2.62) and (2.63). The charges within the cell at R' create a potential

$$\sum_L \frac{\sqrt{4\pi}}{2l+1}\left(\frac{r_R}{w}\right)^l Y_L(\hat{r}_R)\sum_{L'} M_{RL,R'L'}Q_{R'L'}\frac{1}{2w}, \quad (4.35)$$

around site R. In this expression we have introduced the multipole moments defined for the Wigner–Seitz cell as

$$Q_{RL} = \frac{\sqrt{4\pi}}{2l+1}\int_{\Omega_R}\left(\frac{r_R}{w}\right)^l n_R(\mathbf{r})Y_L(\hat{r}_R)d\mathbf{r}_R - Z_R\delta_{L,L_0}, \quad (4.36)$$

which by means of the shape function may be rewritten as

$$Q_{RL} = \frac{\sqrt{4\pi}}{2l+1}\int_0^{s_R^c}\left(\frac{r_R}{w}\right)^l \tilde{n}_{RL}(r_R)r_R^2 dr_R - Z_R\delta_{L,L_0}. \quad (4.37)$$

Now, integrating the potential (4.35) multiplied by the density within Ω_R and taking into account the nuclear charge Z_R within this cell, we arrive at the following expression for the interaction energy between charges within $\Omega_{R'}$ and Ω_R

$$\frac{1}{2w}\sum_{LL'} Q_{RL}M_{RLR'L'}Q_{R'L'}, \quad (4.38)$$

where $M_{RLR'L'}$ is the Madelung matrix introduced in (2.65). Summing up for all R and $R' \neq R$ with non-overlapping bounding spheres, for the inter-cell energy we obtain

$$F_{inter}^{no}[n] = \frac{1}{2w}\sum_{RR'}{}' \sum_{LL'} Q_{RL} M_{RLR'L'} Q_{R'L'}. \tag{4.39}$$

This expression correctly describes the interaction energy between cells with non-overlapping bounding spheres.

4.2.5 Electrostatic Interaction of Neighboring Cells

Equation (4.35) gives the electrostatic potential created by the charge distribution within $\Omega_{R'}$ at a point $\mathbf{r} = \mathbf{r_R} + \mathbf{R}$ outside of the sphere circumscribed to the cell at R', *i.e.* for any $|\mathbf{r} - \mathbf{R}'| \geq s_{R'}^c$. However, in the case of neighboring cells in the region between $\Omega_{R'}$ and its circumscribed sphere we have $|\mathbf{r} - \mathbf{R}'| < s_{R'}^c$. Therefore, Equation (4.38) is no longer valid for such a pair of cells. Mathematically this means that in expansion

$$\frac{1}{|\mathbf{r}' - \mathbf{r}|} = 4\pi \sum_L \frac{1}{2l+1} \frac{r_R^l}{|\mathbf{r}_{R'} + \mathbf{d}_{RR'}|^{l+1}} Y_L(\hat{\mathbf{r}}_R) Y_L(\widehat{\mathbf{r}_{R'} + \mathbf{d}_{RR'}}) \tag{4.40}$$

the necessary condition for the convergence, *i.e.* $r_R < |\mathbf{r}_{R'} + \mathbf{d}_{RR'}|$, is not fulfilled everywhere. Here we have introduced the notation $\mathbf{d}_{RR'} \equiv \mathbf{R}' - \mathbf{R}$.

In fact, it is found that the total inter-cell energy, if calculated using Equation (4.38) for both the non-overlapping and overlapping cells, diverges with increasing l. In order to illustrate this, in Figure 4.5 we show the total electrostatic energy of an *fcc* lattice with a uniform charge distribution. The energy was computed from Equations (4.31) and (4.39) and the overlap between the central site and its 12 nearest-neighbor sites was neglected. The electrostatic energy is scaled so as to yield the Madelung constant of the lattice defined as

$$\alpha_M = -E_c/(Z^2/w). \tag{4.41}$$

Within the Atomic Sphere Approximation (ASA) this equals 1.8, while the exact value for an *fcc* lattice obtained using the Ewald technique [39] is 1.79174723. In this test, the l-truncation was set to the same l_{max}^m for both the intra-cell and inter-cell energy. As we will see later, an l_{max}^m =10−12 gives a well converged result for the intra-cell energy. Therefore, the trend from the figure reflects the divergence of the Madelung energy for the overlapping cells calculated according to Equation (4.38).

Several methods have been proposed to treat this problem [95, 97, 99, 100, 101]. Here we follow the approach introduced by Gonis *et al.* [95] and implemented by Vitos and Kollár [94]. This approach is based on shifting and back-shifting the neighboring cells R' and R with a displacement vector $b_{RR'}$.

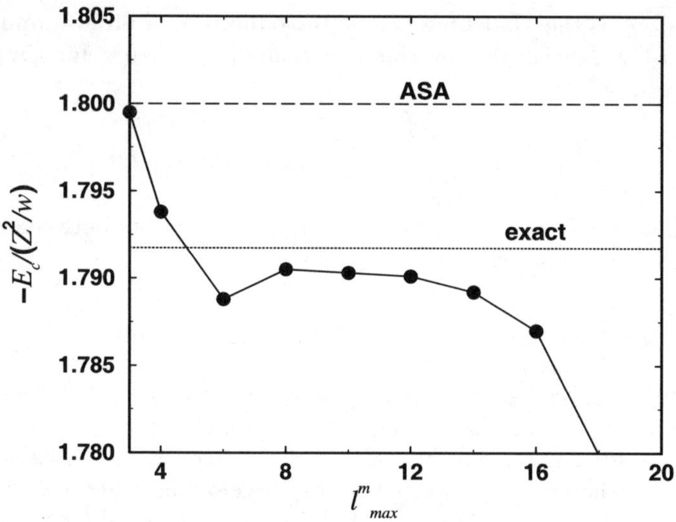

Fig. 4.5. Convergence test for the electrostatic energy of an fcc lattice with homogeneous charge distribution. The total electrostatic energy is plotted as a function of the maximum l used in Equations (4.31) and (4.39) and it is scaled so as to yield the Madelung constant of the lattice.

One can always find a "small" vector $\mathbf{b}_{RR'}$ such that $|\mathbf{r}_{R'}+\mathbf{d}_{RR'}+\mathbf{b}_{RR'}-\mathbf{r}_R| > b_{RR'}$. In this case, we can write

$$\frac{1}{|\mathbf{r}'-\mathbf{r}|} = 4\pi \sum_L \frac{b^l_{RR'}}{2l+1} Y_L(\hat{b}_{RR'})$$
$$\times \frac{1}{|\mathbf{r}'_{R'}+\mathbf{d}_{RR'}+\mathbf{b}_{RR'}-\mathbf{r}_R|^{l+1}} Y_L(\widehat{\mathbf{r}'_{R'}+\mathbf{d}_{RR'}+\mathbf{b}_{RR'}}-\mathbf{r}_R). \quad (4.42)$$

In this expansion the l sum is always convergent. Next we ask that $b_{RR'}$ should be large enough to remove the overlap between the bounding spheres. Assuming that the direction of the displacement vector coincides with the direction between the two cells, the above condition may be expressed as

$$d_{RR'} + b_{RR'} > s^c_R + s^c_{R'}. \quad (4.43)$$

In this situation, for any r_R and $r_{R'}$ we have $r_R < |\mathbf{r}_{R'}+\mathbf{d}_{RR'}+\mathbf{b}_{RR'}|$ and thus the right hand side of Equation (4.42) can be re-expanded around $r^{l'}_R Y_{L'}(\hat{r}_R)$. The coefficients of this second expansion are proportional to $1/|\mathbf{d}_{RR'}+\mathbf{b}_{RR'}|^{l''+1} Y_{L''}(\widehat{\mathbf{d}_{RR'}+\mathbf{b}_{RR'}})$. This can be expanded again around $r^{l'''}_{R'} Y_{L'''}(\hat{r}_{R'})$. After summing up for all the pairs of cells having overlapping bounding spheres, we arrive at [44, 94]

$$F^{ov}_{inter}[n] = \frac{1}{2w}\sum_{RR'}{}' \sum_{L} \frac{1}{2l+1}\left(\frac{b_{RR'}}{w}\right)^{l} Y_L(\hat{b}_{RR'}) \sum_{L',L''} Q_{RL'}$$
$$\times \frac{4\pi(2l''-1)!!}{(2l-1)!!(2l'-1)!!} C_{LL'L''}\delta_{l'',l+l'} \sum_{L'''} M_{RL''\tilde{R}'L'''} Q_{R'L'''}, \qquad (4.44)$$

where $\tilde{\mathbf{R}}' \equiv \mathbf{R}' + \mathbf{b}_{RR'}$ denotes the position of the displaced cell. We have shown that, as in the case of non-overlapping cells, the interaction energy between overlapping cells can be expressed in terms of the original multipole moments and the Madelung matrix. However, in the latter case, the Madelung matrix is calculated for the displaced lattice sites. The summations in Equation (4.44) are conditionally convergent and their range of convergence depends sensitively on the choice of $b_{RR'}$ [94, 95].

Displacement Vector $b_{RR'}$
In order to find a reasonable choice for the displacement vector, we calculate the electrostatic interaction energy of two truncated spheres of radius $s^c_R = s^c_{R'} = s_c$ with a uniform charge distribution n_0, separated by a distance $d_{RR'}$. This arrangement is illustrated in Figure 4.5 for neighboring spheres with $d_{RR'} < s^c_R + s^c_{R'}$ (upper panel) and for a distance $d'_{RR'} > s^c_R + s^c_{R'}$, when the "bounding spheres" do not overlap (lower panel).

In the case of overlapping bounding spheres, the interaction energy for the two truncated spheres is calculated from Equation (4.44) by shifting one of the spheres by $b_{RR'}$. This is given by

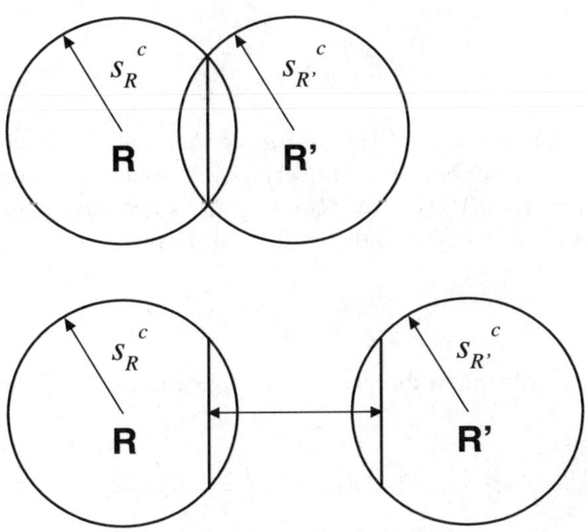

Fig. 4.6. Two truncated spheres of radii s^c_R and $s^c_{R'}$ separated by a distance $|\mathbf{R}'-\mathbf{R}|$ for neighboring case (upper panel) and for non-overlapping "bounding spheres" case (lower panel).

$$\sum_l \left(\frac{b}{b+d}\right)^l \sum_{l',l''} Q_{l'} \frac{(l+l'+l'')!}{l!l'!l''!} \frac{1}{(b+d)^{l'+l''+1}} Q_{l''}, \tag{4.45}$$

where for simplicity we have introduced the notations $d = d_{RR'}$ and $b = b_{RR'}$. For non-overlapping spheres at a minimum distance of $d' \approx 2s^c$, the interaction energy is obtained from (4.38) as

$$\sum_{l,l'} Q_l \frac{(l+l')!}{l!l'!} \frac{1}{(d')^{l+l'+1}} Q_{l'}. \tag{4.46}$$

The multipole moments of the truncated spheres are

$$Q_l = \frac{2\pi n_0}{l+3} {s_c}^{l+3} \int_{-1}^{\frac{d}{2s_c}} P_l(\mathrm{x})d\mathrm{x} + \frac{2\pi n_0}{l+3} \int_{\frac{R}{2s_c}}^{1} \left(\frac{d}{2x}\right)^{l+3} P_l(\mathrm{x})d\mathrm{x}. \tag{4.47}$$

Here $P_l(x)$s are the Legendre polynomials [102]. We mention that owing to the axial symmetry of the system the multiple moments vanish for $m \neq 0$.

The necessary condition for the convergency of the outer sum in Equation (4.45) is the convergency of the inner sums over l' and l'' for each value of l. The diagonal terms of the inner sum in (4.45),

$$A_{l'l'} = {Q_{l'}}^2 \frac{(l+2l')!}{l!(l'!)^2} \frac{1}{(b+d)^{2l'+1}}, \tag{4.48}$$

for $l, l' \gg 1$ have a maximum around $l' = \tilde{l}$, where

$$\tilde{l}(l) = l \frac{\frac{Q_{\tilde{l}+1}}{Q_{\tilde{l}}}}{b+d-2\frac{Q_{\tilde{l}+1}}{Q_{\tilde{l}}}}. \tag{4.49}$$

The upper limit for the ratio of the multipole moments $\frac{Q_{l+1}}{Q_l}$ can be estimated from the interaction energy of non-overlapping spheres. Using the fact that the summation in (4.46) is always convergent, since the inequality (4.43) is already fulfilled for $b = 0$, for high l values we have

$$\frac{Q_{l+1}}{Q_l} < \frac{1}{2} d' \approx s^c. \tag{4.50}$$

The same result is obtained directly from Equation (4.47). Therefore, for $\tilde{l}$ we obtain

$$\tilde{l}(l) < \alpha l \quad \text{with} \quad \alpha \equiv \frac{1}{2} \frac{2s^c}{b+d-2s^c}. \tag{4.51}$$

Thus, we see that the individual terms in the inner sum show a maximum, which (strictly speaking its upper bound) is proportional to l and the coefficient α depends on the displacement vector. In order to ensure the required accuracy of the inner sum it is reasonable to assume that the summation should be carried out at least up to $\simeq 2\alpha l > 2\tilde{l}(l)$ for any l. It is easy to show that in this case for the neglected terms in the inner summations $\frac{A_{l'l'}}{A_{\tilde{l}\tilde{l}}} < 1\%$. This should hold for the largest value of l as well, *i.e.* we obtain the general relation

$$l_{max}^{inner} \simeq 2\alpha l_{max}^{outer}. \tag{4.52}$$

In order to ensure similar accuracy for different neighbors we have to choose $\alpha = const.$ which leads to the equation

$$b + d = \left(1 + \frac{1}{2\alpha}\right) 2s^c \tag{4.53}$$

or more generally

$$b_{RR'} + d_{RR'} = \left(1 + \frac{1}{2\alpha}\right) (s_R^c + s_{R'}^c). \tag{4.54}$$

This equation serves as a general recipe to choose the displacement vector $b_{RR'}$ for an arbitrary pair of overlapping cells. We can see that the limit $\alpha \to \infty$ corresponds to the lower limit of $b_{RR'} + d_{RR'}$ in the inequality (4.43), but according to (4.52) in this case the inner summation should go to infinity. For finite l_{max}^{inner}, $b_{RR'} + d_{RR'}$ has to be chosen according to (4.54) above its lower limit to assure the convergency of the inner summation for any l.

4.2.6 The l Summations

In the following, we discuss the convergence properties of different quantities that appear in the expression of the FCD total energy (4.23). First of all, we should point out the distinction between the maximum ls used in the EMTO and FCD formalisms. The number of exact muffin-tin orbitals is l_{max} and it is usually set to 3. The total number of tail functions or *highers* (see Section 2.1.2) is l_{max}^h. Usually, this parameter represents the maximum l included in the one-center expansion of the charge density (see Section 2.2) and its actual value depends on how anisotropic the charge distributions within the Wigner–Seitz cells are. Within the FCD scheme, the l-truncation used for the shape function plays the central role. As we pointed out earlier, the partial components with $l > l_{max}^s$ are usually neglected in Equation (4.5). Therefore, it is l_{max}^s that fixes the maximum ls used in Equations (4.29), (4.31) and (4.39) and the maximum ls for the inner summations in Equation (4.44).

Charge Neutrality

During the self-consistent EMTO calculation, after each iteration the Fermi level is re-adjusted through Equation (2.44) so that the condition (2.45) is always exactly fulfilled. Here the electron count is realized via the overlap matrix (2.43), and thus via the energy derivative of the kink matrix. On the other hand, in the total energy calculation based on the FCD formalism, the total number of electrons is found directly by integrating the electron density (2.47) within the unit cell. In an ideal situation, this integral should give $N_e = \sum_R Z_R$. However, due to the numerical approximation, in practice there is always a charge misfit

$$\begin{aligned} \Delta_e(l^h_{max}) &= \sum_R \int_{\Omega_R} n(\mathbf{r}_R) d\mathbf{r}_R - N_e \\ &= \sum_R \sqrt{4\pi} \int_0^{s^c_R} \tilde{n}_{RL_0}(r_R) r_R^2 dr_R - N_e. \end{aligned} \tag{4.55}$$

Since, according to Equation (4.6),

$$\sqrt{4\pi}\tilde{n}_{RL_0}(r_R) = \sum_L n_{RL}(r_R)\sigma_{RL}(r_R) \tag{4.56}$$

and usually $l^h_{max} \ll l^s_{max}$, the charge misfit depends on l^h_{max} rather than on l^s_{max}. The same is true for the the kinetic energy given by Equation (4.24) together with (4.27). We mention that the requirement $\Delta_e \to 0$ is one of the most severe tests not only for the one-center expansion but also for the accuracy of the slope matrix and its energy derivative.

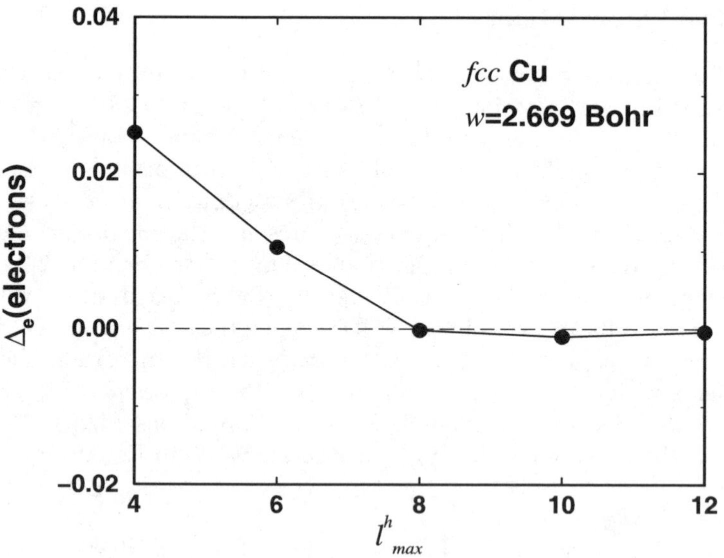

Fig. 4.7. The calculated charge misfit Δ_e for fcc Cu plotted as a function of l^h_{max}.

In Figure 4.7, $\Delta_e(l^h_{max})$ is shown as a function of l^h_{max} in the case of fcc Cu. We can see that even in this relatively symmetric case, an $l^h_{max} \geq 8$ is needed to decrease the error in the total number of electrons within the Wigner–Seitz below ~ 0.001 electrons. In the following tests, we assume that such a convergence of the one-center expansion of the charge density is assured.

Intra-cell Hartree Energy
The convergence properties of the intra-cell electrostatic energy have been studied in detail [44, 48]. In Figure 4.8, the intra-cell Hartree energy of fcc Cu is plotted relative to its converged value as a function of l^s_{max} used in (4.31). As may be seen from the figure, the energy difference of ~ 0.3 mRy, obtained for $l^s_{max} = 8$, is reduced below 0.1 mRy for $l^s_{max} = 12$ and below a few μRy for $l^s_{max} \geq 20$. This result holds for other more open structures as well. We have found that for a wide range of crystal structures a convergence better than 10 μRy of the intra-cell energy can be achieved by performing the summation over l in Equation (4.31) up to 14–20.

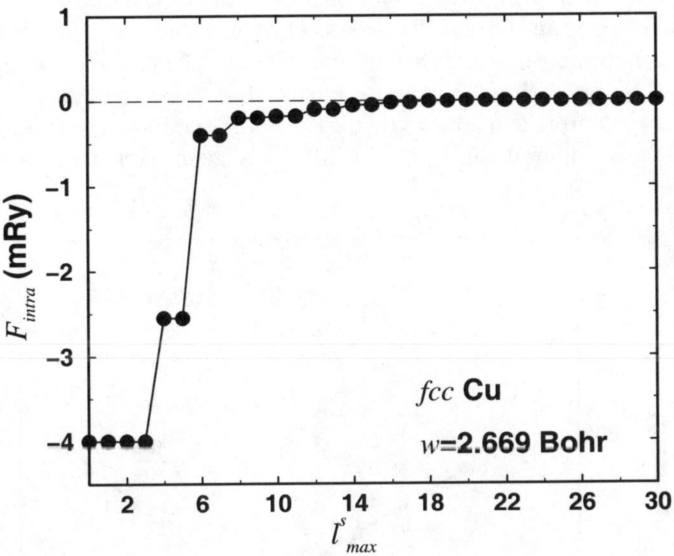

Fig. 4.8. Convergence test for the intra-cell Hartree energy of fcc Cu. The results are plotted relative to their converged value as a function of the maximal l value used in Equation (4.31).

Madelung Energy
Actual calculations of the interaction energy in Equation (4.39) for cell with non-overlapping bounding spheres show that $\sim$ 10 μRy accuracy can be achieved by performing the summation over l and l' up to 8–14 for a wide

range of structures. On the other hand, as we will see later, an accurate evaluation of the interaction energy for overlapping cells (Equation (4.44)) requires multipole moments with $l > 14$. Because of this, the summation in Equation (4.39) is carried out up to the same l as the one used for the inner summations from (4.44). In practice, the Madelung matrix is generated for $l^m_{max} = 6 - 8$ using the Ewald technique, and for $l > 8$ it is calculated on a real space cluster consisting of the 60−80 nearest-neighbor lattice sites.

Next we turn to the inter-cell energy for cells with overlapping bounding spheres and investigate how the inner and outer summations from Equation (4.44) are connected. The first question that arises, is how to choose the parameter α in Equation (4.54). For this we again make use of the model system introduced in Section 4.2.5. In Figure 4.9, we plotted the maximal l values of the outer (l^{outer}_{max}) and the inner (l^{inner}_{max}) sums, which are required to achieve a given accuracy ($\sim 0.01\%$) of the Coulomb energy (4.45) as functions of α. Note that because of the multipole moments, in fact l^{inner}_{max} represents l^s_{max}. As one can see from the figure, a maximal l^{outer}_{max} used in the calculation defines a lower limit, while a maximal l^{inner}_{max} an upper limit for the allowed α values. With increasing α, the number of terms in the outer sum and thus the necessary computer time decreases. On the other hand, with increasing α, we need more and more multipole moments in the calculation of the inner sum. Since the most time consuming part of the calculation of the multipole moments is the determination of the partial components of the shape function, which can be tabulated once and for all for a given structure, we prefer to

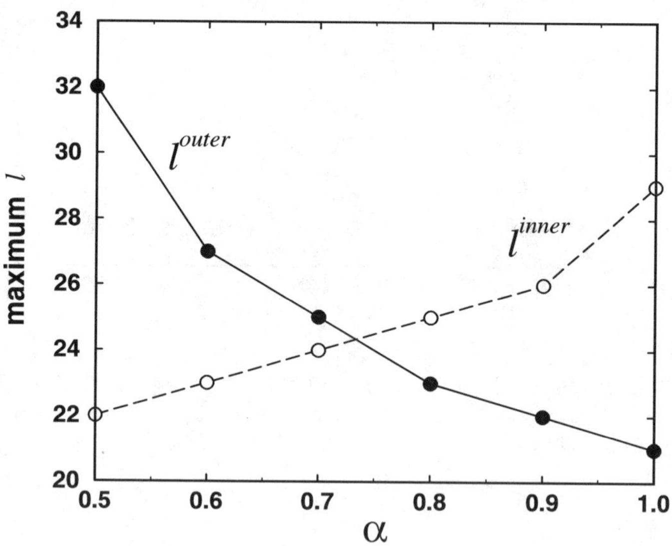

Fig. 4.9. The l^{outer}_{max} and l^{inner}_{max} used in the outer and inner sums as functions of α. For each α the accuracy of the Coulomb energy achieved is $\sim 0.01\%$.

Table 4.2. The Madelung constant for *bcc*, *fcc*, *sc* hexagonal close-packed (*hcp*), simple tetragonal (*tet*) with $c/a = 1.5$ and α–Pu crystal structures.

	bcc	*fcc*	*sc*	*hcp*	*tet*	α-Pu
α_M	1.79185851	1.79174723	1.76011888	1.79167624	1.65374661	1.76614295

minimize the computer time needed for the summations and chose α near its upper limit. In practice, we have found that $\alpha \approx 1$, $l^{inner}_{max} = l^{s}_{max} = 28{-}30$ and l^{outer}_{max} =18–22 lead to an optimal choice for both the accuracy and computer time.

The validity of the choice given in Equation (4.54) for $b_{RR'}$ should hold for realistic calculations as well, because in Equation (4.50) for the ratio of the multipole moments we use an estimation which is independent of the shape of the cell and of the charge distribution. In order to test this, we apply

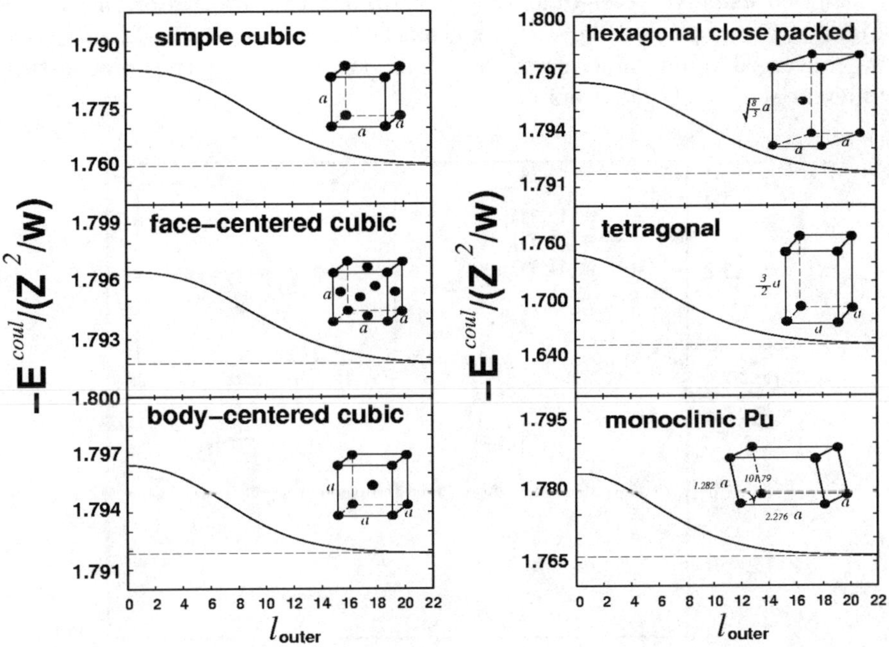

Fig. 4.10. The scaled average Coulomb energy per cell of a homogeneous charge distribution (Madelung constant) of simple, face-centered and body-centered cubic (left panel) and hexagonal close-packed, tetragonal and α-Pu (right panel) structures as a function of l^{outer}_{max}. For the tetragonal lattice $c/a = 1.5$ was used. For α-Pu in the schematic plot of the unit cells the atoms within the cells are not shown. The exact values for the Madelung constants obtained by the Ewald procedure are indicated by dashed lines.

the method to calculate the electrostatic energy of a homogeneous charge distribution of several lattices with different symmetry.

For a few selected structures, the average Madelung constants calculated using the Ewald technique are listed in Table 4.2. For the same structures, the scaled Coulomb energies calculated using the FCD formalism are plotted in Figure 4.10 as a function of l_{max}^{outer}. In these tests, for the intra-cell electrostatic energy given in (4.31) and for the inter-cell electrostatic energy for cells with non-overlapping bounding spheres (4.39) the converged values were used for any l_{max}^{outer}. We can see from the figure that the results converge to the exact values smoothly in each case, indicating that using (4.54) the sums over l converge simultaneously for different neighboring cells. The relative deviations from the exact values are less than 0.03% for $l_{max}^{outer} = 20$ in each case.

Exchange-correlation Energy

Here we discuss the convergence properties of the exchange-correlation energy term, calculated from Equation (4.29). The surface integral over θ and ϕ is performed using the two-dimensional (2D) Gaussian integration method. In Figure 4.11, we plotted the exchange-correlation energy of fcc Cu, relative to its converged value, in terms of l_{max}^s. Different symbols correspond to three different sets of $2D$ mesh points.

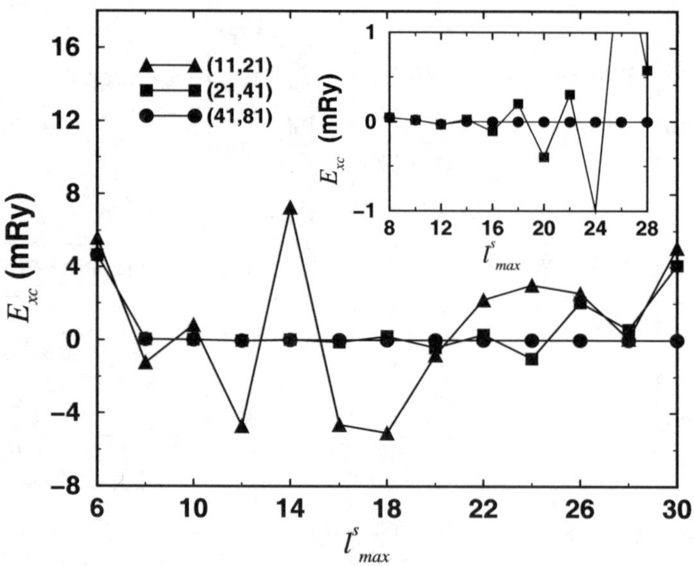

Fig. 4.11. Convergence test for the exchange-correlation energy of fcc Cu as function of the maximal l values used in Equation (4.29). The energies are plotted relative to the converged result. The numbers in parenthesis denote the total number of θ and ϕ Gaussian mesh points on the spherical surface.

It is seen on the figure that no convergence can be achieved for a small number of points ($N_\theta = 11, N_\phi = 21$). By doubling the number of 2D mesh points the converged value is recovered for l^s_{max} =8–10, but for $l^s_{max} > 16-18$ the energy starts to oscillate and it diverges. Only for a very large number of mesh points does the summation in Equation (4.29) become absolutely convergent. This behavior is connected with the fact that for large l values, which are important for the proper mapping of the shape of the Wigner–Seitz cell (especially in the case of open structures), the spherical harmonics have more and more structure, and this cannot be described correctly unless the surface integral is carried out with very high accuracy.

5

The EMTO-CPA Method

In Chapter 1, we briefly presented the most important aspects of the commonly used approaches to describe the energetics of fully or partially disordered solids (Section 1.3). Among these, the most powerful technique in the case of multicomponent alloys is the *Coherent Potential Approximation* (CPA). In this chapter, first we shall outline the main features of the CPA, and then present its implementation within the Exact Muffin-Tin Orbitals formalism. Since the algebraic formulation of the EMTO-CPA method is very similar to that of the EMTO method, here we shall concentrate only on those details where the extension is not straightforward. At the end of the chapter, the main difference between the EMTO-CPA method and former CPA methods will be discussed.

5.1 Coherent Potential Approximation

The *Coherent Potential Approximation* was introduced by Soven [68] for the electronic structure problem and by Taylor [69] for phonons in random alloys. Later, Györffy [70] formulated the CPA in the framework of the multiple scattering theory using the Green function technique. The CPA is based on the assumption that the alloy may be replaced by an ordered effective medium, the parameters of which are determined self-consistently. The impurity problem is treated within the *single-site* approximation. This means that one single impurity is placed in an effective medium and no information is provided about the individual potential and charge density beyond the sphere or polyhedra around this impurity. Below, we illustrate the principal idea of the CPA within the conventional muffin-tin formalism.

We consider a substitutional alloy $A_aB_bC_c...$, where the atoms A, B, C, ... are randomly distributed on the underlying crystal structure. Here $a, b, c, ...$ stand for the atomic fractions of the A, B, C, ... atoms, respectively. This system is characterized by the Green function g and the alloy potential P_{alloy}. The latter, due to the environment, shows small variations around the same

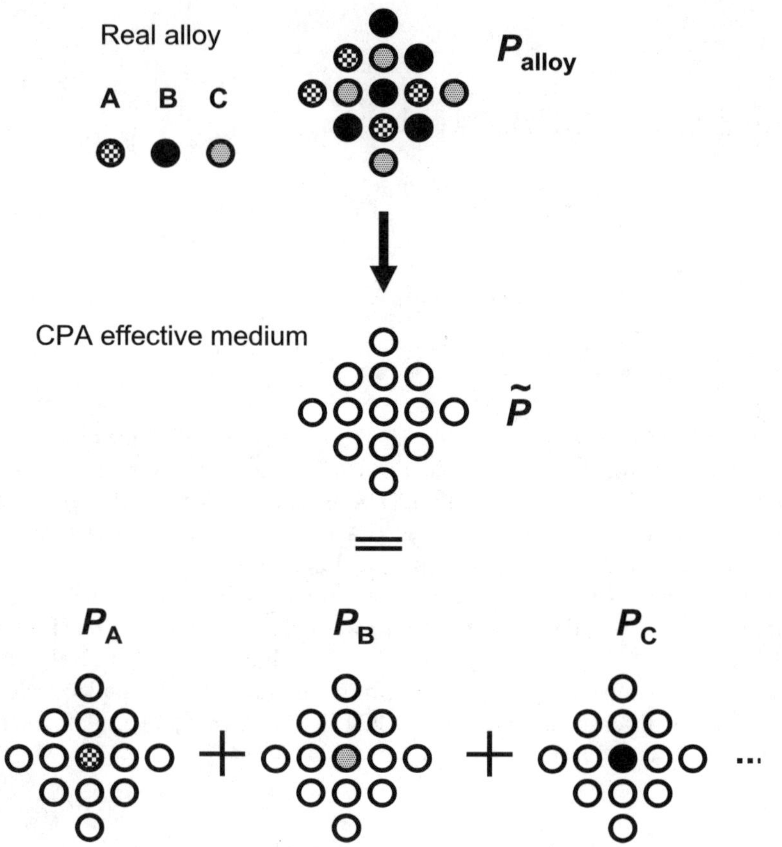

Fig. 5.1. Illustration of the *Coherent Potential Approximation* to the alloy problem. The real alloy, composed by atoms A, B, C, ..., within the CPA is replaced by an effective medium. Given are the notations for the potentials: P_{alloy} is the real alloy potential, $\tilde{P}$ is the coherent potential, $P_{\text{A}}, P_{\text{B}}, P_{\text{C}}, ...$ are the potentials of the alloy components.

type of atoms. There are two main approximations within the CPA. First, it is assumed that the local potentials around a certain type of atom from the alloy are the same, *i.e.* the effect of local environments is neglected. These local potentials are described by the potential functions[1] $P_{\text{A}}, P_{\text{B}}, P_{\text{C}}, ...$ Second, the system is replaced by a monoatomic set-up described by the site independent

[1] For the definition of the potential function within the muffin-tin formalism see [20, 39]

coherent potential $\tilde{P}$. In terms of Green functions, one approximates the real Green function g by a *coherent Green function* $\tilde{g}$. For each alloy component i =A, B, C,... a single-site Green function g_i is introduced. A schematic plot of this idea is given in Figure 5.1.

The main steps to construct the CPA effective medium are as follows. First, the coherent Green function is calculated from the coherent potential using an electronic structure method. Within the Korringa–Kohn–Rostoker (KKR) [36, 37, 51, 52, 103, 104] or Linear Muffin-Tin Orbital (LMTO) [20, 39, 40] methods, we have

$$\tilde{g} = \left[S - \tilde{P}\right]^{-1}, \tag{5.1}$$

where S denotes the KKR or LMTO structure constant matrix corresponding to the underlying lattice. Next, the Green functions of the alloy components, g_i, are determined by substituting the coherent potential of the CPA medium by the real atomic potentials P_i. Mathematically, this condition is expressed via the real-space Dyson equation

$$g_i = \tilde{g} + \tilde{g}\left(P_i - \tilde{P}\right) g_i \quad , \quad i = \mathrm{A, B, C...} \tag{5.2}$$

Finally, the average of the individual Green functions should reproduce the single-site part of the coherent Green function, *i.e.*

$$\tilde{g} = a g_{\mathrm{A}} + b g_{\mathrm{B}} + c g_{\mathrm{C}} + ... \tag{5.3}$$

Equations (5.1)–(5.3) are solved iteratively, and the output $\tilde{g}$ and g_is are used to determine the electronic structure, charge density and total energy of the random alloy.

Nowadays, the CPA has become a state-of-the-art technique for electronic structure calculations in random alloys. Numerous applications [77, 105, 106, 107, 108, 109, 110, 111, 112, 113, 114, 115, 116, 117, 118] have shown that within this approximation one can calculate lattice parameter, bulk modulus, mixing enthalpy, *etc.*, with an accuracy similar to that obtained for ordered solids. At the same time, the CPA, being a single-site approximation to the impurity problem, has limited applicability. For example, one cannot take into account directly within the CPA the effect of short-range order. Also, systems with a large size mismatch between the alloy components are difficult to describe because of the local lattice relaxations.

The most spectacular failure of the existing CPA methods happens in the case of anisotropic lattice distortions in random alloys. This problem was also attributed to the inherent single-site approximation. However, one should bear in mind that certain limitations of the CPA are not directly related to the approximation itself. Rather, they originate from additional approximations introduced by particular implementations. The most common electronic structure calculations methods used for the implementations of the CPA are the

KKR and the LMTO methods based on the Atomic Sphere Approximation (ASA). The shape approximation for the electron density and potential, used in these methods, is insufficient for the accurate description of the behavior of the total energy upon anisotropic lattice distortions. Thus, one cannot calculate, for example, elastic constants in random alloys or relax c/a ratio in alloys with a tetragonal or hexagonal symmetry. In addition, the LMTO-ASA method does not give a proper description of the open structures or structural energy differences between structures with different packing fractions, to the extent that even the energy difference between the *bcc* and *fcc* structures of late transition metals is incorrectly described [119]. However, the most recent reformulation of the CPA [78, 79], demonstrates that this approximation implemented within the framework of the EMTO theory, in contrast to the traditional KKR-CPA and LMTO-CPA methods, is suitable to reproduce the structural energy differences and energy changes related to small lattice distortions in random alloys with high accuracy.

5.2 Fundamentals of the EMTO-CPA Method

The EMTO theory formulates an efficient and at the same time accurate muffin-tin method for solving the Kohn–Sham equations of the Density Functional Theory. By using large overlapping potential spheres, the EMTO approach describes more accurately the exact crystal potential than any conventional muffin-tin method. Furthermore, in the latter methods, the shape approximation used for the potential and density is carried on to the solution of the one-electron equations as well. In contrast to this, in the EMTO approach, while keeping the simplicity and efficiency of the muffin-tin formalism, the one-electron states are calculated exactly for the model potential. This is why the EMTO theory provides an ideal ground for developing an accurate and efficient CPA related method for random alloys.

5.2.1 Average EMTO-CPA Green Function

We consider a substitutional alloy with a fixed underlying lattice. We denote the unit cell sites of the underlying lattice by Q, Q', etc. On each site Q, we have N_Q alloy components. The atomic fractions of the components give the concentrations c_Q^i ($i = 1, 2, ..., N_Q$). The individual spherical potentials are denoted by $v_Q^i(r_Q)$ and they are defined within the potential spheres of radii s_Q^i. We note that these potentials are somewhat different from the spherical potentials $v_R^i(r_R)$ defined in the case of the real alloy because of different local environments. Within the CPA, however, we make the approximation [120]

$$v_Q^i(r_Q) \approx v_R^i(r_R). \tag{5.4}$$

Accordingly, all the potential dependent functions, such as the partial waves, logarithmic derivatives, normalization functions, *etc.*, belonging to the same sort of atom but different Rs are also assumed to be the same. Because of this, in the following, we use index R when we refer to the real space and index Q when we refer to the quantities written for the underlying lattice.

For each sort of atom, the partial waves are constructed from the solutions $\phi^i_{Rl}(\epsilon, r_R)$ of the radial Schrödinger equation (2.20). From the matching conditions at s^i_R and the radius a_R of the corresponding a-sphere[2], we set up the backwards extrapolated free-electron solutions $\varphi^i_{Rl}(\epsilon, r_R)$, the logarithmic derivative $D^i_{Rl}(\epsilon)$ and the normalization $N^i_{Rl}(\epsilon)$ functions. For the sake of simplicity, in this chapter we omit the screening index a. Formally, the above functions are obtained from Equations (2.22), (2.26) and (2.27), by substituting $D^a_{Rl}(\epsilon)$ with $D^i_{Rl}(\epsilon)$ and $\phi_{Rl}(\epsilon, s_R)$ with $\phi^i_{Rl}(\epsilon, s_R)$.

The CPA effective medium is described by a site (Q) dependent coherent potential, which possesses the symmetry of the underlying crystal lattice. In EMTO formalism, the coherent potential is introduced via the logarithmic derivative $\tilde{D}_{QL'QL}(z)$ of the effective scatterers, and, therefore, the coherent Green function or the path operator (5.1) is given by [78]

$$\sum_{Q''L''} a_{Q'} \left[S_{Q'L'Q''L''}(\kappa^2, \mathbf{k}) - \delta_{Q'Q''} \tilde{D}_{Q'L'Q'L''}(z) \right] \times \tilde{g}_{Q''L''QL}(z, \mathbf{k}) = \delta_{Q'Q} \delta_{L'L}, \tag{5.5}$$

where $l, l', l'' \leq l_{max}$, and $S_{Q'L'Q''L''}(\kappa^2, \mathbf{k})$ are the elements of the EMTO slope matrix for complex energy $\kappa^2 = z - v_0$ and Bloch vector $\mathbf{k}$ from the Brillouin zone (BZ). For ordered systems, this equation reduces to Equation (2.34). The logarithmic derivative of the effective scatterers is site-diagonal with non-zero $L' \neq L$ off-diagonal elements.

The average of the on-site (QQ) elements of the Green function for alloy component i, $g^i_{QLQL'}$, is calculated as an impurity Green function of the ith alloy component embedded in the effective medium. In the single-site approximation, this is obtained from the real space Dyson equation (5.2) as a single-site perturbation on the coherent potential as

$$g^i_{QLQL'}(z) = \tilde{g}_{QLQL'}(z) + \sum_{L''L'''} \tilde{g}_{QLQL''}(z) \times \left[D^i_{Ql''}(z) \delta_{L''L'''} - \tilde{D}_{QL''QL'''}(z) \right] g^i_{QL'''QL'}(z), \tag{5.6}$$

where $D^i_{Ql}(z) \approx D^i_{Rl}(z)$ is the logarithmic derivative function for the ith alloy component and

$$\tilde{g}_{QLQL'}(z) = \int_{BZ} \tilde{g}_{Q''L''QL}(z, \mathbf{k}) d\mathbf{k} \tag{5.7}$$

[2] Note that the radii of the hard spheres are determined by the underlying lattice and they do not depend on i.

is the site-diagonal part of the k-integrated coherent Green function. The condition of vanishing scattering, on the average, leads to relation (5.3) between $\tilde{g}_{QLQL'}(z)$ and the Green functions of alloy components, namely

$$\tilde{g}_{QLQL'}(z) = \sum_i c_Q^i \, g_{QLQL'}^i(z), \tag{5.8}$$

Equations (5.5),(5.6) and (5.8) are solved self-consistently for $\tilde{D}(z), \tilde{g}(z, \mathbf{k})$ and $g^i(z)$. The total number of states below the Fermi level,

$$N(\epsilon_F) = \frac{1}{2\pi i} \oint_{\epsilon_F} < G(z) > dz, \tag{5.9}$$

is obtained from the average Green function

$$< G(z) > \equiv \int_{BZ} \sum_{Q'L'QL} \tilde{g}_{Q'L'QL}(z, \mathbf{k}) \, a_Q \dot{S}_{QLQ'L'}(\kappa^2, \mathbf{k}) d\mathbf{k} \; -$$

$$- \sum_{Qi} c_Q^i \sum_L \left[g_{QLQL}^i(z) \, a_Q \dot{D}_{Ql}^i(z) + \left(\frac{\dot{D}_{Ql}^i(z)}{D_{Ql}^i(z)} - \frac{1}{z - e_{Ql}^i} \right) \right], \tag{5.10}$$

where the *overdot* stands for the energy derivative, and $l, l' \leq l_{max}$. The site off-diagonal elements of the coherent Green function $\tilde{g}_{Q'L'QL}(z, \mathbf{k})$ are calculated from Equation (5.5) with the self-consistent logarithmic derivative $\tilde{D}_{QL'QL}(z)$ of the effective scatterers. For ordered systems, the above expressions reduce to Equations (2.45) and (2.46) from Section 2.1.5.

The first term from the right hand side of Equation (5.10) assures the proper normalization of the one-electron states for the optimized overlapping potential. In fact, within the single-site approximation for the impurity Green function, Equation (5.9) gives the exact number of states at the Fermi level [78, 79].

5.2.2 Full Charge Density

The complete non-spherically symmetric charge density of alloy component i is represented in one-center form

$$n_R^i(\mathbf{r_R}) = \sum_L n_{RL}^i(r_R) Y_L(\hat{r}_R). \tag{5.11}$$

This expansion is obtained from the real-space expression $< G(z, \mathbf{r}_R, \mathbf{r}_R) >= \sum_i c^i G^i(z, \mathbf{r}_R, \mathbf{r}_R)$ of the average Green function (5.10). To this end, we transform the first term from the right hand side of (5.10), the so called interstitial term, to one-center form. Formally, this procedure is the same as that from Section 2.2.

Inside the Wigner–Seitz cell at R, the partial components $n_{RL}^i(r_R)$ of the average density $n_R^i(\mathbf{r}_R)$ belonging to the sublattice Q are determined from the restricted average of the on-site element of the Green function for the ith alloy component $g^i_{QLQL'}$ and from the coherent Green function $\tilde{g}_{QLQ'L'}$. This leads to

$$n^i_{RL}(r_R) = \frac{1}{2\pi i}\oint_{\epsilon_F}\sum_{L''L'} C_{LL'L''} Z^i_{Rl''}(z, r_R) \times \tilde{\mathcal{D}}^i_{RL''L'}(z)\, Z^i_{Rl'}(z, r_R) dz, \tag{5.12}$$

where $C_{LL'L''}$ are the real Gaunt numbers (Appendix B) and $Z^i_{Rl}(z, r_R)$ denote the $Y_L(\hat{r}_R)$ projections of the exact muffin-tin orbitals

$$Z^i_{Rl}(z, r_R) = \begin{cases} N^i_{Rl}(z)\phi^i_{Rl}(z, r_R) & \text{if } l \le l_{max} \text{ and } r_R \le s^i_R \\ \varphi^i_{Rl}(z, r_R) & \text{if } l \le l_{max} \text{ and } r_R > s^i_R \\ -j_l(\kappa\, r_R) & \text{if } l > l_{max} \text{ for all } r_R \end{cases}. \tag{5.13}$$

The low-l block of the density matrix $\mathcal{D}^i$ of the alloy component i is given by

$$\mathcal{D}^i_{RL'L}(z) = g^i_{QL'QL}(z) + \frac{\delta_{L'L}}{a_Q\, \dot{D}^i_{Ql}(z)}\left(\frac{\ddot{D}^i_{Ql}(z)}{D^i_{Ql}(z)} - \sum_{\epsilon^i_{Ql}}\frac{1}{z - \epsilon^i_{Ql}}\right), \tag{5.14}$$

with $l, l' \le l_{max}$. Similarly to Equation (2.51), the second term from the right hand side of (5.14) is introduced to remove the nonphysical poles of the normalization function $N^i_{Rl}(z)$. The high-l–low-l blocks of $\mathcal{D}^i$ are

$$\mathcal{D}^i_{RL'L}(z) = \sum_{Q''L''}\int_{BZ} \tilde{g}_{QL'Q''L''}(z, \mathbf{k}) a_{Q''} S_{Q''L''QL}(\kappa^2, \mathbf{k}) d\mathbf{k} \tag{5.15}$$

for $l' \le l_{max}$ and $l > l_{max}$, and

$$\mathcal{D}^i_{RL'L}(z) = \sum_{Q''L''}\int_{BZ} S_{QL'Q''L''}(\kappa^2, \mathbf{k}) \tilde{g}_{Q''L''QL}(z, \mathbf{k}) d\mathbf{k} \tag{5.16}$$

for $l' > l_{max}$ and $l \le l_{max}$. Finally, the high-l block of $\mathcal{D}^i$ is

$$\mathcal{D}^i_{RL'L}(z) = \sum_{Q''L''R'''L'''}\int_{BZ} S_{QL'Q''L''}(\kappa^2, \mathbf{k}) \tilde{g}_{Q''L''Q'''L'''}(z, \mathbf{k}) \times a_{Q'''} S_{Q'''L'''QL}(\kappa^2, \mathbf{k}) d\mathbf{k}, \tag{5.17}$$

with $l', l > l_{max}$. Note that the expressions for the high-l–low-l and high-l–high-l blocks of $\mathcal{D}^i$ do not involve the Green functions for the alloy component.

5.2.3 The EMTO-CPA Effective Potential

In the case of alloys, we define an overlapping muffin-tin potential $v^i_{mt}(\mathbf{r})$ (see Equation (2.1)), for each alloy component i at sublattice Q of the underlying crystal lattice. The spherical potential $v^i_R(r_R)$ is calculated from the restricted average Green function for the ith alloy component, following the main idea of the CPA as an impurity embedded in the effective medium. The electrostatic potential of the electronic and protonic charge densities is divided into components due to the charges from inside and from outside of the potential sphere. The spherical potential due to the charges inside of the potential sphere is given by (see Equation (2.60))

$$\begin{aligned} v^{I,i}_R(r_R) &= 8\pi \frac{1}{r_R} \int_0^{r_R} {r'_R}^2 n^i_{RL_0}(r'_R) dr'_R \\ &+ 8\pi \int_{r_R}^{s^i_R} r'_R n^i_{RL_0}(r'_R) dr'_R - \frac{2Z^i_R}{r_R}, \end{aligned} \tag{5.18}$$

where Z^i are the protonic charges and $L_0 \equiv (0,0)$. The net charges from the outside of the potential sphere are taken into account by the average Madelung potential from Equation (2.64), where Q^{SCA}_{RL} are the average multipole moments calculated within the spherical cells,

$$\begin{aligned} Q^{SCA}_{RL} &= \sum_i c^i_R \Big[\frac{\sqrt{4\pi}}{2l+1} \int_0^{w^i_R} \left(\frac{r_R}{w}\right)^l n^i_{RL}(r_R)\, r^2_R\, dr_R \\ &- Z^i_R\, \delta_{L,L_0} \Big] + \delta^{SCA}\, \delta_{L,L_0}. \end{aligned} \tag{5.19}$$

w^i_R denotes the radius of the spherical cell of type i at site R. The site independent normalization constant δ^{SCA} is determined from the condition of charge neutrality $\sum_R Q^{SCA}_{RL_0} = 0$.

Due to the SCA (see Section 2.3.1), the average number of electrons inside the s-spheres at R,

$$Q^s_R = \sum_i c^i_R\, Q^{i,s}_R = \sum_i c^i_R \sqrt{4\pi} \int_0^{s^i_R} r^2_R n^i_{RL_0}(r_R) dr_R, \tag{5.20}$$

is different from the average number of electrons inside the cell, which contributes with a constant shift to the spherical potential. Within the EMTO-CPA method, the SCA shift is calculated from Equation (2.68) with

$$\Delta Q_{R_{NN}} \equiv \frac{1}{N_{NN}} \left(Q^{SCA}_{RL_0} + \sum_i c^i_R Z^i_R - Q^s_R \right), \tag{5.21}$$

where N_{NN} is the number of nearest-neighbor cells, and $Q^{SCA}_{RL_0}$ and Q^s_R are given by (5.19) and (5.20), respectively.

Since the impurity problem is treated within the single-site approximation, the Coulomb system of a particular alloy component may contain a non-zero net charge. The effect of the charge misfit on the spherical potential is taken into account using the screened impurity model (SIM) by Korzhavyi *et al.* [121, 122]. According to this model, an additional shift of

$$\Delta v_R^{SIM,i} = -\frac{2\alpha_c}{w}\left(Q_R^{i,s} - Q_R^s\right), \tag{5.22}$$

is added to the spherical part of the full-potential around site R. Here, $Q_R^{i,s}$ and Q_R^s are defined in (5.20). Note that in the case of ordered systems, $\Delta v_R^{CPA,i} = 0$. In Equation (5.22), α_c is a dimensionless parameter that controls the effectiveness of screening around the impurity. We shall return to this parameter at the end of Section 5.2.4.

The total potential within the potential sphere of the ith alloy component is obtained as the sum of the intracell potential (5.18), the Madelung potential (2.64), the SCA (2.68) and SIM (5.22) corrections, and the spherical symmetric exchange-correlation potential (2.69) calculated from the density of the ith alloy component, namely

$$v_R^i(r_R) = v_R^{I,i}(r_R) + v_R^M + \Delta v_R^{SCA} + \Delta v_R^{SIM,i} + \mu_{xcR}^i(r_R). \tag{5.23}$$

Finally, the interstitial potential within the SCA is obtained according to (2.56) as the average interstitial potential calculated from $v_R^i(r_R)$

$$v_0 = \sum_{Ri} c_R^i \int_{s_R^i}^{w_R^i} r_R^2\, v_R^i(r_R)\, dr_R / \sum_{Ri} c_R^i \left[({w^i_R}^3 - {s^i_R}^3)/3\right]. \tag{5.24}$$

5.2.4 The EMTO-CPA Total Energy

The total energy of the random alloy is calculated according to

$$\begin{aligned} E_{tot} &= T_s[n] + \sum_R \sum_i c^i \left(F_{intraR}^i[n_R^i] + E_{xcR}^i[n_R^i]\right) + \\ &+ F_{inter}[Q] + \Delta E^{SIM}. \end{aligned} \tag{5.25}$$

The kinetic energy is determined from the one-electron equations,

$$\begin{aligned} T_s[n] &= \frac{1}{2\pi i}\oint_{\epsilon_F} z < G(z) > dz \\ &- \sum_R \sum_i c^i \int_{\Omega_R} v_{mt}^i(\mathbf{r}_R) n_R^i(\mathbf{r}_R) d\mathbf{r}_R, \end{aligned} \tag{5.26}$$

where the first term from the right hand side is the sum of the average one-electron energies with $< G(z) >$ from Equation (5.10) and $v^i_{mt}(\mathbf{r}_R)$ is the optimized overlapping muffin-tin potential for sort i constructed from (5.23) and (5.24).

The Coulomb energy components of the total energy functional are calculated in a similar way to the EMTO-FCD method (Sections 4.2.3, 4.2.4 and 4.2.5). In particular, the EMTO-CPA intra-cell energy is

$$\begin{aligned} F^i_{intraR}[n^i_R] &= \frac{\sqrt{4\pi}}{w} \sum_L \int_0^{s^c_R} \tilde{n}^i_{RL}(r_R) \left[\left(\frac{r_R}{w}\right)^l P^i_{RL}(r_R) \right. \\ &+ \left. \left(\frac{r_R}{w}\right)^{-l-1} Q^i_{RL}(r_R) - 2Z^i_R \frac{w}{r_R} \delta_{L,L_0} \right] r_R^2 dr_R \end{aligned} \tag{5.27}$$

where $\tilde{n}^i_{RL}(r_R)$ is calculated from (4.6) using for the density of the alloy component i,

$$P^i_{RL}(r_R) \equiv \frac{\sqrt{4\pi}}{2l+1} \int_{r_R}^{s^c_R} \tilde{n}^i_{RL}(r'_R) \left(\frac{r'_R}{w}\right)^{-l-1} (r'_R)^2 dr'_R, \tag{5.28}$$

and

$$Q^i_{RL}(r_R) \equiv \frac{\sqrt{4\pi}}{2l+1} \int_0^{r_R} \tilde{n}^i_{RL}(r'_R) \left(\frac{r'_R}{w}\right)^{l} (r'_R)^2 dr'_R. \tag{5.29}$$

The EMTO-CPA inter-cell energy is calculated from Equations (4.39) and (4.44) using for the multipole moments the average moments

$$Q_{RL} = \sum_i c^i_R Q^i_{RL}, \tag{5.30}$$

with

$$Q^i_{RL} = \frac{\sqrt{4\pi}}{2l+1} \int_0^{s^c_R} \left(\frac{r_R}{w}\right)^l \tilde{n}^i_{RL}(r_R) r_R^2 dr_R - Z^i_R \delta_{L,L_0}. \tag{5.31}$$

Finally, the EMTO-CPA exchange-correlation energy is

$$\begin{aligned} E_{xc}[n_R] &= \int_0^{2\pi} \int_0^{\pi} \int_0^{s^c_R} n^i_R(\mathbf{r}_R) \varepsilon_{xc}([n^i_R]; \mathbf{r}_R) \\ &\times \sum_L^{l^s_{max}} \sigma_{RL}(r_R) Y_L(\hat{r}_R)\, r_R^2 \, dr_R \, \sin\theta d\theta d\phi. \end{aligned} \tag{5.32}$$

For notation and convergence properties, see Sections 4.1, 4.2.2 and 4.2.6.

The last term from (5.25) is the screened impurity model correction [121, 122] to the electrostatic energy given by

$$\Delta E^{SIM} = -\sum_i c^i \frac{\alpha'_c}{w} \left(Q_R^{i,s} - Q_R^s \right)^2 . \tag{5.33}$$

The SIM parameter α'_c is determined from the condition that the total energy (5.25) calculated within the CPA should match the total energy of the real alloy calculated using a large unit cell [122]. For most of the alloys, this is achieved for $\alpha'_c \approx 0.6 - 1.0$. In many applications, α_c from (5.22) and α'_c from (5.33) are chosen to be the same. When the two SIM parameters are treated as independent parameters, α_c can be used, *e.g.*, to match the charge transfer obtained within the CPA and the real charge transfer.

5.3 EMTO-CPA Method versus Other CPA Methods

The main difference between the EMTO expression for the average Green function and the one defined within former CPA methods, such as the KKR-CPA or LMTO-CPA methods, is featured in Figure 5.2. Circles stand for the muffin-tin or atomic spheres centered on lattice sites and the shaded area is the region between the spheres.

In order to understand this difference, we decompose the EMTO-CPA Green function (5.10) into a single-site contribution G^i_{ss} and multi-site contribution $\tilde{G}_{\text{I}}$, *viz.*

$$< G > = \sum_i c^i G^i_{\text{ss}} + \tilde{G}_{\text{I}} . \tag{5.34}$$

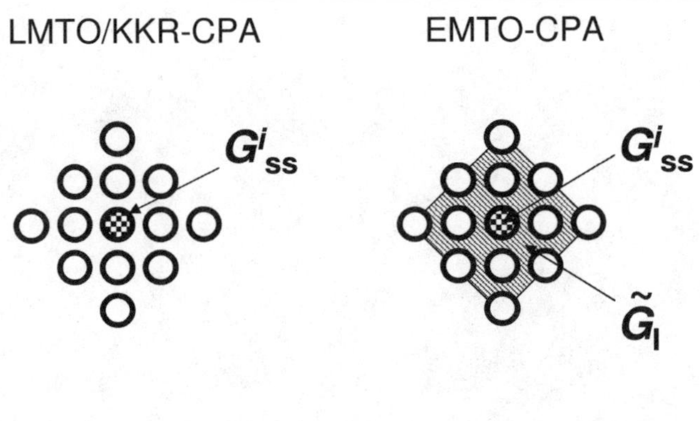

Fig. 5.2. Illustration of the difference between conventional muffin-tin based CPA Green functions (KKR-CPA, LMTO-CPA) and the EMTO-CPA Green function, Equations (5.10) or (5.34). Shaded area is the interstitial region defined as the domain outside of the muffin-tin spheres.

The single-site Green function $G^i_{\rm ss}$ involves the Green functions g^i and the logarithmic derivatives D^i of the alloy components [120], whereas $\tilde{G}_{\rm I}$ depends only on the coherent Green function $\tilde{g}$. The latter is attributed to several lattice sites rather than to a single-site. It corresponds approximately to the interstitial states and assures a proper normalization of the single-electron states [46]. This term was ignored (actually suppressed into the single-site term $G^i_{\rm ss}$) in the former implementations of the CPA. It is important to note that within the single-site approximation, the Green function from Equation (5.34) leads to the exact density of states for the overlapping muffin-tin potential $v^i_{mt}(\mathbf{r}_R)$.

Part II

Applications

6

Ground-state Properties

In this chapter, we shall review some important ground-state properties that can be derived directly from the total energy calculated for different volumes, geometries and compositions. For a specific space group, chemical composition and magnetic structure, the equation of state is obtained from the total energy $E(V)$ calculated as a function of volume (V). At each volume, the crystal structure should be fully relaxed with respect to the internal coordinates and unit cell shape. The negative volume derivative of $E(V)$ gives the pressure,

$$P(V) = -\frac{\partial E(V)}{\partial V}. \tag{6.1}$$

From the pressure–volume relation, we get the enthalpy $H(P) = E(V(P)) + PV(P)$. The bulk modulus is defined from the volume derivative of the pressure as

$$B(V) = -V\frac{\partial P}{\partial V} = V\frac{\partial^2 E(V)}{\partial V^2}. \tag{6.2}$$

The third order volume derivative of $E(V)$ enters the expression of the pressure derivative of the bulk modulus (B') or the Grüneisen constant[1]. In order to minimize the numerical noise in P, B and B', usually an analytic form is used to fit the *ab initio* energies versus volume. In Section 6.1, we shall describe the most commonly used equation of states: the Murnaghan, the Birch–Murnaghan and the Morse functions.

The elastic constants may be computed from the strain derivative of the total energy. In Section 6.2, we shall briefly review the theory of elastic constants and present the technique used to obtain the single crystal elastic constant in cubic and hexagonal lattices. The polycrystalline elastic moduli will be introduced in Section 6.3. In Section 6.4, the theoretical determination of the

[1] The Grüneisen constant (γ) describes the anharmonic effects in the vibrating lattice and it is given by $\gamma = -f + B'/2$. The best agreement with the experimental Grüneisen constant was obtained [123] for $f = 1/2$.

surface energy, surface stress and stacking fault energy will be presented. A few ideas about the atomistic modeling of the mechanical properties of alloys will be discussed in Section 6.6. Finally, in Section 6.7 we shall list some useful hints for total energy calculations using the EMTO method.

6.1 Equation of State

The Murnaghan equation of state [124, 125] originates from the observation that the pressure derivative of the bulk modulus shows negligible pressure dependence. Therefore, we can make the approximation

$$\frac{\partial B}{\partial P} \approx B_0' = \left.\frac{\partial B}{\partial P}\right|_{V=V_0}, \tag{6.3}$$

where V_0 is the equilibrium volume. Using the definition of the bulk modulus, from Equation (6.3) we obtain

$$B_0' = -\frac{V}{B}\frac{\partial B}{\partial V}. \tag{6.4}$$

After integrating between V_0 and V, this leads to

$$B(V) = B_0 \left(\frac{V_0}{V}\right)^{B_0'}. \tag{6.5}$$

Repeated integrating gives the pressure

$$P(V) = \frac{B_0}{B_0'}\left[\left(\frac{V_0}{V}\right)^{B_0'} - 1\right], \tag{6.6}$$

and the total energy

$$E(V) = E_0 + \frac{B_0 V}{B_0'}\left(\frac{(V_0/V)^{B_0'}}{B_0' - 1} + 1\right) - \frac{B_0 V_0}{B_0' - 1}, \tag{6.7}$$

where E_0 is the energy at the equilibrium volume. A somewhat more flexible equation of state is the third-order Birch–Murnaghan function given by [126, 127]

$$E(V) = E_0 + \frac{9}{16} B_0 V_0 \times \left\{ \left[\left(\frac{V_0}{V}\right)^{\frac{2}{3}} - 1 \right]^3 B_0' - \left[\left(\frac{V_0}{V}\right)^{\frac{2}{3}} - 1 \right]^2 \left[4 \left(\frac{V_0}{V}\right)^{\frac{2}{3}} - 6 \right] \right\}, \quad (6.8)$$

and

$$P(V) = \frac{3}{2} B_0 \left[\left(\frac{V_0}{V}\right)^{\frac{7}{3}} - \left(\frac{V_0}{V}\right)^{\frac{5}{3}} \right] \left\{ 1 + \frac{3}{4}(B_0' - 4) \left[\left(\frac{V_0}{V}\right)^{\frac{2}{3}} - 1 \right] \right\}. \quad (6.9)$$

The bulk modulus can be obtained from Equation (6.9) according to Equation (6.2). Both the Murnaghan and Birch–Murnaghan functions involve four independent parameters: E_0, V_0, B_0 and B_0'. While E_0, V_0, B_0 vary significantly from one material to another, the value of B_0' is fairly constant for many substances [128].

Sometimes it is more preferable to work with a Morse type of equation of state. The total energy is fitted by an exponential function [123]

$$E(w) = a + be^{-\lambda w} + ce^{-2\lambda w} \quad (6.10)$$

written in terms of the average Wigner–Seitz radius w. Here λ, a, b and c are the four independent Morse parameters. Since $(4\pi w^2)\partial/\partial V = \partial/\partial w$, the expression for the pressure becomes

$$P(w) = \frac{x\lambda^3}{4\pi(\ln x)^2}(b + 2cx), \quad (6.11)$$

where

$$x \equiv e^{-\lambda w}. \quad (6.12)$$

The equilibrium Wigner–Seitz radius, defined by $V_0 = 4\pi w_0^3/3$, is obtained from the condition $P(w_0) = 0$ as

$$w_0 = -\frac{\ln x_0}{\lambda} \text{ with } x_0 = -\frac{b}{2c}. \quad (6.13)$$

From Equation (6.11), we get the bulk modulus as a function of the Wigner–Seitz radius as

$$B(w) = -\frac{x\lambda^3}{12\pi \ln x}\left[(b + 4cx) - \frac{2}{\ln x}(b + 2cx)\right], \quad (6.14)$$

which at w_0 reduces to

$$B_0 = -\frac{cx_0^2\lambda^3}{6\pi \ln x_0}. \quad (6.15)$$

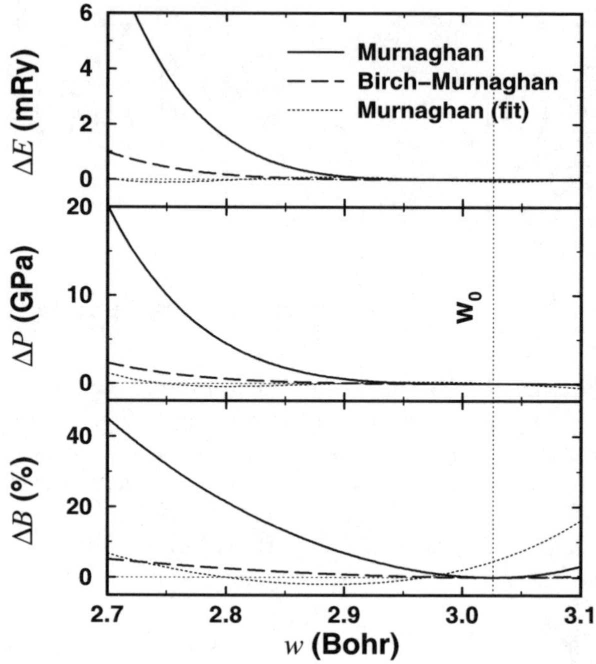

Fig. 6.1. The difference between total energy (upper panel, in mRy), pressure (middle panel, in GPa) and bulk modulus (lower panel, in percent) for *fcc* Ag obtained from the Murnaghan and Birch–Murnaghan equations of states relative to the Morse equation of state. The fitting parameters for Murnaghan (solid line) and Birch–Murnaghan (dashed line) are chosen to be identical with those obtained from the Morse fit. The dotted line is a fit to the *ab initio* total energies using the Murnaghan function.

Finally, the pressure derivative of the bulk modulus at V_0 becomes

$$B_0' = 1 - \ln x_0. \tag{6.16}$$

Equations (6.13), (6.15), (6.16) and the expression for the equilibrium energy $E_0 = a + bx_0 + cx_0^2$ give the connection between the Morse fitting parameters λ, a, b and c and the parameters E_0, B_0, V_0 and B_0' used in the Murnaghan or Birch–Murnaghan equation of states.

In Figure 6.1, we compare the three equations of states for *fcc* Ag. The parameters $E_0 = -10626.843688$ Ry, $w_0 = 3.026$ Bohr, $B_0 = 104.4$ GPa and $B_0' = 5.81$ were obtained by fitting the Morse function to 7 *ab initio* energies calculated for atomic radii between 2.7 and 3.1 Bohr. In these calculations the Local Airy Approximation [17, 18] was employed for the exchange-correlation functional. For comparison, the experimental equilibrium atomic radius and bulk modulus are 3.018 Bohr and 98.8 GPa [129], and from the experimental Grüneisen constant 2.4 [123] we get $B_0' = 5.92$. The above theoretical Morse parameters were used for all three equations of states from Figure 6.1. The agreement between the total energy, pressure and bulk modulus obtained by the Birch–Murnaghan (dashed line) and Morse functions is satisfactory. However, near $w = 2.7$ Bohr (corresponding to $P \approx 95$ GPa) the pressure and bulk modulus obtained from the Murnaghan expression (solid line) differ by about 20% and 40%, respectively, from those obtained by the other two expressions. On the other hand, fitting the Murnaghan function to the *ab initio* energies (dotted line), we obtain a much better agreement. The so obtained parameters are $E_0 = -10626.843742$ Ry, $w_0 = 3.027$ Bohr, $B_0 = 110.2$ GPa and $B_0' = 4.79$. The improved agreement between the Morse and Murnaghan equation of states is mainly due to the $\sim 18\%$ decrease in B_0'.

6.2 Single Crystal Elastic Constants

The adiabatic[2] elastic constants are the second order derivatives of the internal energy with respect to the strain tensor e_{kl} $(k, l = 1, 2, 3)$, *viz.*

$$c_{ijkl} = \frac{1}{V}\frac{\partial E}{\partial e_{ij} \partial e_{kl}}, \tag{6.17}$$

where E stands for the internal energy and the derivatives are calculated at constant entropy and constant es other than e_{ij} and e_{kl}. The elastic constants form a fourth-rank tensor, which can be arranged in a 6×6 matrix with maximum 21 different elements. Employing the Voigt notation[3], the elastic compliances $s_{\alpha\beta}$ are related to the elastic constants by

$$\sum_{\gamma} c_{\alpha\gamma} s_{\gamma\beta} = \delta_{\alpha\beta}, \tag{6.18}$$

where $\delta_{\alpha\beta}$ is the Kronecker *delta*. The elastic constants determine, for example, the acoustic velocities v through the Christoffel equation [54]

$$det(c_{ijkl} n_j n_k - \rho v^2 \delta_{il}) = 0, \tag{6.19}$$

where n is the propagation direction and ρ is the density. There are usually two quasi-transversal and one quasi-longitudinal real roots. The ratio between extremal propagation directions defines the acoustic anisotropy.

---

[2] The isothermal elastic constants are obtained as the second order strain derivative of the Helmholtz free energy calculated at constant temperature. For temperatures below the Debye temperature, there is no significant difference between the adiabatic and isothermal elastic constants [128].

[3] In the Voigt notation the pair of indices ij are replaced by index α according to: $\alpha = 1, 2, 3$ for $ij = 11, 22, 33$, $\alpha = 4$ for $ij = 23$ or 32, $\alpha = 5$ for $ij = 13$ or 31 and $\alpha = 6$ for $ij = 12$ or 21.

In the following, we concentrate on lattices with cubic and hexagonal symmetries. In a cubic lattice, there are three independent elastic constant c_{11}, c_{12} and c_{44}. Then, the anisotropic variation of the sound velocity is determined by the parameter

$$A_{\mathrm{E}} = \frac{c_{11} - c_{12} - 2c_{44}}{c_{11} - c_{44}} \tag{6.20}$$

introduced by Every [130]. An alternative measure of anisotropy is the Zener ratio [131]

$$A_{\mathrm{Z}} = \frac{2c_{44}}{c_{11} - c_{12}}. \tag{6.21}$$

For an isotropic cubic crystal $c_{11} - c_{12} = 2c_{44}$, so $A_{\mathrm{E}} = 0$ and $A_{\mathrm{Z}} = 1$. In a hexagonal lattice, there are five independent elastic constants $c_{11}, c_{12}, c_{13}, c_{33}$ and c_{44}, and the anisotropy is described in terms of one compressional

$$\Delta_P = \frac{c_{33}}{c_{11}}, \tag{6.22}$$

and two shear

$$\Delta_{S1} = \frac{c_{11} + c_{33} - 2c_{13}}{4c_{44}} \quad \text{and} \quad \Delta_{S2} = \frac{2c_{44}}{c_{11} - c_{12}} \tag{6.23}$$

anisotropy ratios. Hence, the hexagonal lattice is isotropic if $c_{11} = c_{33}, c_{12} = c_{13}$ and $c_{11} - c_{12} = 2c_{44}$.

The dynamical or mechanical stability condition of a lattice implies that the energy change $\Delta E \sim V c_{\alpha\beta} e_\alpha e_\beta$ upon any small deformation is positive. This condition can be formulated in terms of elastic constants [128]. The stability criteria for cubic crystals requires that

$$c_{44} > 0 \ , \ c_{11} > |c_{12}| \ , \ c_{11} + 2c_{12} > 0. \tag{6.24}$$

For a hexagonal lattice the stability conditions are

$$\begin{aligned} &c_{44} > 0 \ , \ c_{11} > |c_{12}| \ , \ c_{11}\, c_{33} > (c_{13})^2, \\ &c_{33}(c_{11} + c_{12}) > 2(c_{13})^2. \end{aligned} \tag{6.25}$$

From the single crystal elastic constants one obtains the sound velocity $v_s(\theta, \phi)$ for the longitudinal (s = L) and the two transverse (s = $\mathrm{T_1,T_2}$) branches. They are solutions of the Christoffel Equation (6.19). Explicit expressions for $v_s(\theta, \phi)$ in the case of cubic and hexagonal crystals can be found in [128]. The average of $v_s{}^{-3}(\theta, \phi)$ over all directions (θ, ϕ) gives the sound velocity v_m, *viz.*

$$v_m^{-3} = \frac{1}{3} \sum_s \frac{1}{4\pi} \int v_s{}^{-3}(\theta, \phi) \sin\theta d\theta d\phi. \tag{6.26}$$

This velocity is used in the conventional Debye model with the Debye temperature defined as

$$\theta_D = \frac{\hbar}{k_B}\left(\frac{6\pi^2}{V}\right)^{1/3} v_m. \tag{6.27}$$

Here V is the atomic volume, $\hbar$ and k_B are the Planck and Boltzmann constants, respectively[4]. This Debye temperature gives, for instance, the low-temperature limit of the heat capacity per atom $C_D(T) = (12\pi^4/5)k_B(T/\theta_D)^3$, [128]. At high-temperatures, however, the entropy depends on the logarithmic average of all the phonon frequencies rather than on the arithmetic average, Equation (6.26). Then, in a Debye-model description of the entropy, one should use the logarithmic average $v_{\log}$ of the sound velocities $v_s(\theta, \phi)$ and an entropy-related Debye temperature $\theta_{D,\log}$. We note that neither θ_D nor $\theta_{D,\log}$ are defined for systems where the stability requirements, expressed, *e.g.*, via Equations (6.24) and (6.25), are violated.

6.2.1 Numerical Calculation of the Elastic Constants

At volume V, the elastic constants are obtained by straining the lattice and evaluating the total energy changes due to the strain as a function of its magnitude. We choose the applied strains to be volume conserving (except for the bulk modulus). This is important since the total energy depends on the volume much more strongly than on strain. By choosing volume conserving strains we obviate the separation of these two contributions to the total energy. Using isochoric strains we assure the identity of our calculated elastic constants with the stress–strain coefficients, which are appropriate for the calculation of elastic wave velocities; this identity is nontrivial for finite applied pressure. We denote by $e_1, e_2, ..., e_6$ the elements of the stain matrix, *i.e.*

$$\mathcal{D}(e) = \begin{pmatrix} e_1 & \frac{1}{2}e_6 & \frac{1}{2}e_5 \\ \frac{1}{2}e_6 & e_2 & \frac{1}{2}e_4 \\ \frac{1}{2}e_5 & \frac{1}{2}e_4 & e_3 \end{pmatrix}. \tag{6.28}$$

The energy change upon strain (6.28) is written as

$$E(e_1, e_2, ..., e_6) = E(0) + \frac{1}{2}V \sum_{i,j=1,6} c_{ij} e_i e_j + \mathcal{O}(e^3), \tag{6.29}$$

where $E(0)$ is the energy of the undistorted lattice, and $\mathcal{O}(e^3)$ denotes the terms proportional to e^k with $k \geq 3$. An arbitrary vector $\mathbf{r} = (x, y, z)$ under stain (6.28) transforms to $\mathbf{r}' = (x', y', z')$ according to

[4] In atomic Rydberg units, $\hbar = 1$ and $k_B = 6.33363 \times 10^{-6}$ Ry/K.

$$\begin{pmatrix} x' \\ y' \\ z' \end{pmatrix} = (\mathcal{D}(e) + \mathcal{I}) \begin{pmatrix} x \\ y \\ z \end{pmatrix} = \begin{pmatrix} (1+e_1)x + \frac{1}{2}e_6 y + \frac{1}{2}e_5 z \\ \frac{1}{2}e_6 x + (1+e_2)y + \frac{1}{2}e_4 z \\ \frac{1}{2}e_5 x + \frac{1}{2}e_4 y + (1+e_3)z \end{pmatrix}, \quad (6.30)$$

where $\mathcal{I}$ is the 3×3 identity matrix. Mathematically, isochoric strain corresponds to a strain matrix with determinant $det(\mathcal{D} + \mathcal{I}) = 1$.

In practice, $e_1, e_2, ..., e_6$ are expressed as functions of a single parameter δ so that the energy change (6.29) becomes

$$E(\delta) = E(0) + VC\delta^2 + \mathcal{O}(\delta^3), \quad (6.31)$$

where C stands for a particular combination of c_{ij}. We fit $E(\delta)$ by a polynomial of δ, $E(\delta) = E(0) + a_2\delta^2 + a_3\delta^3 + ...$, and C is obtained from the second order coefficient a_2 as[5]

$$C = \frac{a_2}{V}. \quad (6.32)$$

Usually, the total energies are computed for six distortions $\delta = 0.00, 0.01, ..., 0.05$.

Cubic Lattice

In a cubic lattice, having three independent elastic constants, the energy change (6.29) upon stain (6.28) is

$$\frac{1}{V}\Delta E = \frac{1}{2}c_{11}\left(e_1^2 + e_2^2 + e_3^2\right) + c_{12}\left(e_1e_2 + e_2e_3 + e_1e_3\right) + \frac{1}{2}c_{44}\left(e_4^2 + e_5^2 + e_6^2\right) + \mathcal{O}(e^3). \quad (6.33)$$

Two of the cubic elastic constants are derived from the bulk modulus

$$B = \frac{1}{3}(c_{11} + 2c_{12}), \quad (6.34)$$

and the tetragonal shear modulus

$$c' = \frac{1}{2}(c_{11} - c_{12}). \quad (6.35)$$

The bulk modulus corresponds to isotropic lattice expansion and may be obtained from the equation of state (Section 6.1) fitted to the *ab initio* total energies calculated for a set of volumes. The shear modulus c' is obtained from a tetragonal distortion

$$\mathcal{D}_t + \mathcal{I} = \begin{pmatrix} 1+\delta_t & 0 & 0 \\ 0 & 1+\delta_t & 0 \\ 0 & 0 & \frac{1}{(1+\delta_t)^2} \end{pmatrix}, \quad (6.36)$$

[5] The elastic constant are usually expressed in gigapascals, 1 Ry/Bohr$^3 \approx 14710.5$ GPa.

resulting in energy change

$$\Delta E(\delta_\mathrm{t}) = 6Vc'\delta_\mathrm{t}^2 + \mathcal{O}(\delta_\mathrm{t}^3). \tag{6.37}$$

Since this is an asymmetric distortion (odd powers of δ_t appear), calculations have to be carried out for both negative and positive δ_t values. Alternatively, we can use the orthorhombic deformation

$$\mathcal{D}_\mathrm{o} + \mathcal{I} = \begin{pmatrix} 1+\delta_\mathrm{o} & 0 & 0 \\ 0 & 1-\delta_\mathrm{o} & 0 \\ 0 & 0 & \frac{1}{1-\delta_\mathrm{o}^2} \end{pmatrix}, \tag{6.38}$$

which leads to the energy change

$$\Delta E(\delta_\mathrm{o}) = 2Vc'\delta_\mathrm{o}^2 + \mathcal{O}(\delta_\mathrm{o}^4). \tag{6.39}$$

The c_{44} shear modulus may be obtained from the monoclinic distortion

$$\mathcal{D}_\mathrm{m} + \mathcal{I} = \begin{pmatrix} 1 & \delta_\mathrm{m} & 0 \\ \delta_\mathrm{m} & 1 & 0 \\ 0 & 0 & \frac{1}{1-\delta_\mathrm{m}^2} \end{pmatrix}, \tag{6.40}$$

and

$$\Delta E(\delta_\mathrm{m}) = 2Vc_{44}\delta_\mathrm{m}^2 + \mathcal{O}(\delta_\mathrm{m}^4). \tag{6.41}$$

For $\mathcal{D}_\mathrm{o}$ and $\mathcal{D}_\mathrm{m}$, calculations needs to be performed only for positive δ.

To illustrate the above distortions, we consider a cubic lattice with N_Q atoms in the unit cell. The primitive and basis vectors expressed in units of cubic lattice constant are

$$\mathbf{a} = \begin{pmatrix} 1 \\ 0 \\ 0 \end{pmatrix}, \mathbf{b} = \begin{pmatrix} 0 \\ 1 \\ 0 \end{pmatrix}, \mathbf{c} = \begin{pmatrix} 0 \\ 0 \\ 1 \end{pmatrix} \tag{6.42}$$

and

$$\mathbf{Q}_i = \begin{pmatrix} x_i \\ y_i \\ z_i \end{pmatrix} \quad \text{for } i = 1, 2, ..., N_Q. \tag{6.43}$$

For example, let us consider the three primitive cubic lattices. For a simple cubic (sc) lattice $N_Q = 1$, for a body centered cubic (bcc) lattice $N_Q = 2$, and for a face centered cubic (fcc) lattice $N_Q = 4$. The corresponding basis vectors are

$$\mathbf{Q}_1^{sc} = \begin{pmatrix} 0 \\ 0 \\ 0 \end{pmatrix}, \mathbf{Q}_{1,2}^{bcc} = \begin{pmatrix} 0 \\ 0 \\ 0 \end{pmatrix}, \begin{pmatrix} 1/2 \\ 1/2 \\ 1/2 \end{pmatrix}, \tag{6.44}$$

and

$$\mathbf{Q}^{fcc}_{1,2,3,4} = \begin{pmatrix} 0 \\ 0 \\ 0 \end{pmatrix}, \begin{pmatrix} 1/2 \\ 1/2 \\ 0 \end{pmatrix}, \begin{pmatrix} 1/2 \\ 0 \\ 1/2 \end{pmatrix}, \begin{pmatrix} 0 \\ 1/2 \\ 1/2 \end{pmatrix}. \tag{6.45}$$

The tetragonal distortion (6.36) applied on $\mathbf{a}, \mathbf{b}, \mathbf{c}$ and $\mathbf{Q}_i$ given by (6.42) and (6.43) results in

$$\mathbf{a}_\mathrm{t} = \begin{pmatrix} 1+\delta_\mathrm{t} \\ 0 \\ 0 \end{pmatrix}, \mathbf{b}_\mathrm{t} = \begin{pmatrix} 0 \\ 1+\delta_\mathrm{t} \\ 0 \end{pmatrix} \text{ and } \mathbf{c}_\mathrm{t} = \begin{pmatrix} 0 \\ 0 \\ 1/(1+\delta_\mathrm{t})^2 \end{pmatrix}, \tag{6.46}$$

and

$$\mathbf{Q}_{\mathrm{t},i} = \begin{pmatrix} x_i + x_i\delta_\mathrm{t} \\ y_i + y_i\delta_\mathrm{t} \\ z_i/(1+\delta_\mathrm{t})^2 \end{pmatrix}. \tag{6.47}$$

In the case of a *sc* lattice, the resulting structure may be described as a simple tetragonal (*st*), whereas for the *bcc* and *fcc* lattices, the distorted lattice is body centered tetragonal (*bct*). The axial ratios are given in Table 6.1.

The orthorhombic distortion (6.38) applied on $\mathbf{a}, \mathbf{b}, \mathbf{c}$ and $\mathbf{Q}_i$ leads to

$$\mathbf{a}_\mathrm{o} = \begin{pmatrix} 1+\delta_\mathrm{o} \\ 0 \\ 0 \end{pmatrix}, \mathbf{b}_\mathrm{o} = \begin{pmatrix} 0 \\ 1-\delta_\mathrm{o} \\ 0 \end{pmatrix} \text{ and } \mathbf{c}_\mathrm{o} = \begin{pmatrix} 0 \\ 0 \\ 1/(1-\delta_\mathrm{o}^2) \end{pmatrix}, \tag{6.48}$$

and

$$\mathbf{Q}_{\mathrm{o},i} = \begin{pmatrix} x_i + x_i\delta_\mathrm{o} \\ y_i - y_i\delta_\mathrm{o} \\ z_i/(1-\delta_\mathrm{o}^2) \end{pmatrix}. \tag{6.49}$$

For the *sc*, *bcc* and *fcc* parent lattices, the distorted structure may be described as a simple orthorhombic (*so*), a body centered orthorhombic (*bco*) and a face centered orthorhombic (*fco*) lattice, respectively. The orthorhombic lattice constants are listed in Table 6.1.

Finally, the monoclinic distortion (6.40) applied on $\mathbf{a}, \mathbf{b}, \mathbf{c}$ and $\mathbf{Q}_i$ gives the following distorted lattice

$$\mathbf{a}_\mathrm{m} = \begin{pmatrix} 1 \\ \delta_\mathrm{m} \\ 0 \end{pmatrix}, \mathbf{b}_\mathrm{m} = \begin{pmatrix} \delta_\mathrm{m} \\ 1 \\ 0 \end{pmatrix} \text{ and } \mathbf{c}_\mathrm{m} = \begin{pmatrix} 0 \\ 0 \\ 1/(1-\delta_\mathrm{m}^2) \end{pmatrix}, \tag{6.50}$$

Table 6.1. The lattice parameters for the structures obtained by distorting a simple cubic, body centered cubic and face centered cubic lattice using the tetragonal (6.36), orthorhombic (6.38) and monoclinic (6.40) strain matrices.

	sc			*bcc*			*fcc*		
	b/a	c/a	sym.	b/a	c/a	sym.	b/a	c/a	sym.
$\mathcal{D}_{\mathrm{t}}$	1	$\frac{1}{(1+\delta_{\mathrm{t}})^3}$	*st*	1	$\frac{1}{(1+\delta_{\mathrm{t}})^3}$	*bct*	1	$\frac{\sqrt{2}}{(1+\delta_{\mathrm{t}})^3}$	*bct*
$\mathcal{D}_{\mathrm{o}}$	$\frac{1-\delta_{\mathrm{o}}}{1+\delta_{\mathrm{o}}}$	$\frac{1}{(1+\delta_{\mathrm{o}})(1-\delta_{\mathrm{o}}^2)}$	*so*	$\frac{1-\delta_{\mathrm{o}}}{1+\delta_{\mathrm{o}}}$	$\frac{1}{(1+\delta_{\mathrm{o}})(1-\delta_{\mathrm{o}}^2)}$	*bco*	$\frac{1-\delta_{\mathrm{o}}}{1+\delta_{\mathrm{o}}}$	$\frac{1}{(1+\delta_{\mathrm{o}})(1-\delta_{\mathrm{o}}^2)}$	*fco*
$\mathcal{D}_{\mathrm{m}}$	$\frac{1-\delta_{\mathrm{m}}}{1+\delta_{\mathrm{m}}}$	$\frac{1}{\sqrt{2}(1+\delta_{\mathrm{m}})(1-\delta_{\mathrm{m}}^2)}$	*baco*	$\frac{1-\delta_{\mathrm{m}}}{1+\delta_{\mathrm{m}}}$	$\frac{1}{\sqrt{2}(1+\delta_{\mathrm{m}})(1-\delta_{\mathrm{m}}^2)}$	*fco*	$\frac{1-\delta_{\mathrm{m}}}{1+\delta_{\mathrm{m}}}$	$\frac{\sqrt{2}}{(1+\delta_{\mathrm{m}})(1-\delta_{\mathrm{m}}^2)}$	*bco*

and

$$\mathbf{Q}_{\mathrm{m},i} = \begin{pmatrix} x_i + y_i\delta_{\mathrm{m}} \\ x_i\delta_{\mathrm{m}} + y_i \\ z_i/(1-\delta_{\mathrm{m}}^2) \end{pmatrix}. \tag{6.51}$$

For sc, bcc and fcc lattices, the distorted structures are base centered orthorhombic (*baco*), face centered and body centered orthorhombic, respectively. The corresponding axial ratios are given in Table 6.1. Equations (6.46)–(6.51) can be used to establish the primitive cell of the distorted lattice of any cubic lattice (6.42) with basis (6.43).

Hexagonal Lattice
Hexagonal crystals have five independent elastic constants $c_{11}, c_{12}, c_{13}, c_{33}$ and c_{44}, and the energy change (6.29) upon a general stain (6.28) is

$$\frac{1}{V}\Delta E = \frac{1}{2}c_{11}\left(e_1^2+e_2^2\right) + c_{33}e_3^2 + c_{12}e_1e_2 + c_{13}\left(e_2e_3+e_1e_3\right) + \frac{1}{2}c_{44}\left(e_4^2+e_5^2\right) + \frac{1}{2}c_{66}e_6^2 + \mathcal{O}(e^3), \tag{6.52}$$

where $c_{66} = (c_{11}-c_{12})/2$. There are several possibilities to obtain the five elastic constant by calculating ΔE from Equation (6.52) for small distortions [132, 133]. Here we employ the one described by Steinle-Neumann *et al.* [132]. The bulk modulus is calculated from the equation of state as

$$B = \frac{c^2}{c_s}, \tag{6.53}$$

where

$$c^2 \equiv c_{33}(c_{11}+c_{12}) - 2c_{13}^2, \tag{6.54}$$

and

$$c_s \equiv c_{11} + c_{12} + 2c_{33} - 4c_{13}. \tag{6.55}$$

In general the hexagonal axial ratio c/a changes with the volume. Therefore, at each volume V, we have to find the equilibrium structure. This is done by calculating the total energy $E(V, c/a)$ for a series of different c/a ratios and the optimized $(c/a)_0$ is obtained from the minimum of $E(V, c/a)$. The volume dependence of $(c/a)_0(V)$ can be related to the difference in the linear compressibilities along the a and c axes. The dimensionless quantity R describes this as

$$R = -\frac{d\ln(c/a)_0(V)}{d\ln V}, \tag{6.56}$$

which in terms of hexagonal elastic constants becomes

$$R = \frac{c_{33} - c_{11} - c_{12} + c_{13}}{c_s}. \tag{6.57}$$

In some systems, the volume dependence of $(c/a)_0$ is negligible. This is the situation, *e.g.*, in systems with $(c/a)_0$ close to the ideal value $\sqrt{8/3}$. In this case, from Equation (6.57) we have $c_{13} + c_{33} \approx c_{11} + c_{12}$ and the hexagonal bulk modulus reduces to [133]

$$B \approx \frac{2(c_{11} + c_{12}) + 4c_{13} + c_{33}}{9} \approx \frac{2c_{13} + c_{33}}{3}. \tag{6.58}$$

The elastic constant c_s from Equation (6.55) gives the second order energy variation with c/a around the equilibrium value. This is calculated from the isochoric strain

$$\mathcal{D}_\mathrm{h} + \mathcal{I} = \begin{pmatrix} 1+\delta_\mathrm{h} & 0 & 0 \\ 0 & 1+\delta_\mathrm{h} & 0 \\ 0 & 0 & \frac{1}{(1+\delta_\mathrm{h})^2} \end{pmatrix}. \tag{6.59}$$

The corresponding energy change is obtained from Equation (6.52) as

$$\Delta E(\delta_\mathrm{h}) = V c_s \delta_\mathrm{h}^2 + \mathcal{O}(\delta_\mathrm{h}^3). \tag{6.60}$$

From Equations (6.53), (6.55) and (6.57), we obtain the sum of c_{11} and c_{12}. These constants can be separated by computing their difference from the orthorhombic strain

$$\mathcal{D}_\mathrm{o} + \mathcal{I} = \begin{pmatrix} 1+\delta_\mathrm{o} & 0 & 0 \\ 0 & 1-\delta_\mathrm{o} & 0 \\ 0 & 0 & \frac{1}{1-\delta_\mathrm{o}^2} \end{pmatrix}, \tag{6.61}$$

leading to a change in total energy

$$\Delta E(\delta_\mathrm{o}) = 2V c_{66} \delta_\mathrm{o}^2 + \mathcal{O}(\delta_\mathrm{o}^4). \tag{6.62}$$

The fifth elastic constant c_{44} is determined from the monoclinic strain

$$\mathcal{D}_{\mathrm{m}} + \mathcal{I} = \begin{pmatrix} 1 & 0 & \delta_{\mathrm{m}} \\ 0 & \frac{1}{1-\delta_{\mathrm{m}}^2} & 0 \\ \delta_{\mathrm{m}} & 0 & 1 \end{pmatrix}, \tag{6.63}$$

which results in an energy change

$$\Delta E(\delta_{\mathrm{m}}) = 2Vc_{44}\delta_{\mathrm{m}}^2 + \mathcal{O}(\delta_{\mathrm{m}}^4). \tag{6.64}$$

We illustrate the distortions (6.59), (6.61) and (6.63) in the case of a hexagonal close-packed lattice. The corresponding primitive and basis vectors expressed in units of lattice constant a are

$$\mathbf{a} = \begin{pmatrix} 1 \\ 0 \\ 0 \end{pmatrix}, \mathbf{b} = \begin{pmatrix} -1/2 \\ \sqrt{3}/2 \\ 0 \end{pmatrix}, \mathbf{c} = \begin{pmatrix} 0 \\ 0 \\ (c/a)_0 \end{pmatrix} \tag{6.65}$$

and

$$\mathbf{Q}_1 = \begin{pmatrix} 0 \\ 0 \\ 0 \end{pmatrix}, \quad \mathbf{Q}_2 = \begin{pmatrix} 0 \\ \sqrt{3}/3 \\ (c/a)_0/2 \end{pmatrix}, \tag{6.66}$$

where $(c/a)_0$ is the equilibrium axial ratio. The structure obtained from the strain matrix (6.59) remains hexagonal with

$$(c/a)_{\mathrm{h}} = (c/a)_0 \frac{1}{(1+\delta_{\mathrm{h}})^3}. \tag{6.67}$$

Therefore, in Equation (6.60) we can use the same total energies as those used for calculating B and R. First, we compute the total energy $E(V, c/a)$ as a function of V and c/a. Then the bulk modulus is obtained from the second order derivative of the equation of state fitted to the energy minima $E_{min}(V) = \min_{c/a} E(V, c/a) = E(V, (c/a)_0)$, and R is obtained from the optimized $(c/a)_0(V)$ according to Equation (6.56). Finally, the second order derivative of $E(V, c/a)$ near $c/a = (c/a)_0$ gives the c_s, *i.e.*

$$c_s = \frac{9(c/a)_0^2}{V} \left. \frac{\partial^2 E(V, c/a)}{\partial (c/a)^2} \right|_{c/a=(c/a)_0}. \tag{6.68}$$

The strained lattice obtained from the strain matrix (6.61) has the base centered orthorhombic symmetry with two atoms per primitive cell. The lattice parameters and basis vectors in units of the lattice parameter a_{o} are

$$(b/a)_{\mathrm{o}} = \sqrt{3}\frac{1-\delta_{\mathrm{o}}}{1+\delta_{\mathrm{o}}}, \quad (c/a)_{\mathrm{o}} = \frac{(c/a)_0}{(1+\delta_{\mathrm{o}})(1-\delta_{\mathrm{o}}^2)} \tag{6.69}$$

and

$$\mathbf{Q}_{o,1} = \begin{pmatrix} 0 \\ 0 \\ 0 \end{pmatrix}, \quad \mathbf{Q}_{o,2} = \begin{pmatrix} 0 \\ (b/a)_o/3 \\ (c/a)_o/2 \end{pmatrix}. \tag{6.70}$$

The strained lattice obtained from the strain matrix (6.63) is simple monoclinic with four atoms in the primitive cell. The corresponding lattice parameters in units of the lattice parameter a_m are

$$(b/a)_m = (c/a)_o, \quad (c/a)_m = \frac{\sqrt{3}}{\sqrt{1+\delta_m^2(1-\delta_m^2)}}, \tag{6.71}$$

and

$$\cos(\gamma_m) = \frac{2\delta_m}{1+\delta_m^2}, \tag{6.72}$$

where γ_m is the angle between $\mathbf{a}_m$ and $\mathbf{b}_m$. The atomic positions are

$$\mathbf{Q}_{m,1} = \begin{pmatrix} 0 \\ 0 \\ 0 \end{pmatrix}, \mathbf{Q}_{m,2} = \begin{pmatrix} (c/a)_o\delta_m/(1+\delta_m^2) \\ (c/a)_o(1-\delta_m^2)/(1+\delta_m^2)/2 \\ -(c/a)_m/3 \end{pmatrix} \tag{6.73}$$

and

$$\mathbf{Q}_{m,3} = \begin{pmatrix} 1/2 \\ 0 \\ -(c/a)_m/2 \end{pmatrix}, \mathbf{Q}_{m,4} = \mathbf{Q}_{m,2} + \mathbf{Q}_{m,3}. \tag{6.74}$$

Alternatively, this structure may be described as a base centered monoclinic lattice with two atoms per primitive cell [132]. We would like to point out that the above atomic coordinates correspond to a rigidly distorted internal structure. However, the atomic positions from Equations (6.70), (6.73) and (6.74) may vary under strain. When high accuracy is required, the strain relaxation should also be taken into account.

6.3 Polycrystalline Elastic Constants

In the previous section, we showed how the elastic constants of a monocrystalline material can be deduced from *ab initio* total energies calculated for a series of lattices obtained from the parent lattice by applying "small" strains. In a polycrystalline material, the monocrystalline grains are randomly oriented. On a large scale, such materials can be considered to be quasi-isotropic or isotropic in a statistical sense. An isotropic system is completely described by the bulk modulus B and the shear modulus G. The Young modulus E and Poisson ratio ν are connected to B and G by the relations

$$E = \frac{9BG}{3B+G} \quad \text{and} \quad \nu = \frac{3B-2G}{2(3B+G)}. \tag{6.75}$$

In these materials, the sound velocity is isotropic but different for the longitudinal and two transversal branches. The longitudinal velocity is related to the polycrystalline B and G,

$$\rho v_L^2 = B + \frac{4}{3}G, \tag{6.76}$$

and the transversal velocity is related to the polycrystalline G,

$$\rho v_T^2 = G, \tag{6.77}$$

where ρ is the density. Then the average velocity, Equation (6.26), becomes

$$\frac{3}{v_m^3} = \frac{1}{v_L^3} + \frac{2}{v_T^3}, \tag{6.78}$$

which is used in Equation (6.27) to find the polycrystalline Debye temperature.

The only way to establish the *ab initio* polycrystalline elastic moduli is to first derive the monocrystalline elastic constants c_{ij} and then to transform these data to macroscopic quantities by suitable averaging methods based on statistical mechanics. A large variety of methods has been proposed for averaging c_{ij} to obtain the isotropic elastic constants. In the following, we describe the three most widely used averaging methods for the bulk modulus and shear modulus. The corresponding Young modulus and Poisson ratio follow from Equations (6.75).

6.3.1 Averaging Methods

The Voigt and Reuss Bounds

In the Voigt averaging method [128] a uniform strain, and in the Reuss method a uniform stress is assumed. The former is formulated using the elastic constants c_{ij} and the latter using the elastic compliances s_{ij}. Within the Voigt approach, the general expressions for the bulk and shear moduli are

$$B_V = \frac{(c_{11} + c_{22} + c_{33}) + 2(c_{12} + c_{13} + c_{23})}{9}, \tag{6.79}$$

and

$$G_V = \frac{(c_{11} + c_{22} + c_{33}) - (c_{12} + c_{13} + c_{23}) + 3(c_{44} + c_{55} + c_{66})}{15}. \quad (6.80)$$

The corresponding expressions within the Reuss approach are

$$B_R = \frac{1}{(s_{11} + s_{22} + s_{33}) + 2(s_{12} + s_{13} + s_{23})}, \quad (6.81)$$

and

$$G_R = \frac{15}{4(s_{11} + s_{22} + s_{33}) - 4(s_{12} + s_{13} + s_{23}) + 3(s_{44} + s_{55} + s_{66})}, \quad (6.82)$$

respectively.

For cubic crystals, $c_{11} = c_{22} = c_{33}, c_{12} = c_{13} = c_{23}, c_{44} = c_{55} = c_{66}$, and $s_{11} = s_{22} = s_{33}, s_{12} = s_{13} = s_{23}, s_{44} = s_{55} = s_{66}$. Using these relations, for the Voigt and Reuss bounds we obtain

$$\begin{aligned} B_V &= \frac{c_{11} + 2c_{12}}{3}, \\ G_V &= \frac{c_{11} - c_{12} + 3c_{44}}{5}, \end{aligned} \quad (6.83)$$

and

$$\begin{aligned} B_R &= \frac{1}{3\,(s_{11} + 2s_{12})}, \\ G_R &= \frac{5}{4s_{11} - 4s_{12} + 3s_{44}}. \end{aligned} \quad (6.84)$$

Since the cubic elastic compliances may be expressed in terms of the cubic elastic constants (Equation (6.18)) as

$$\begin{aligned} s_{11} + 2s_{12} &= \frac{1}{c_{11} + 2c_{12}}, \\ s_{11} - s_{12} &= \frac{1}{c_{11} - c_{12}}, \\ s_{44} &= \frac{1}{c_{44}}, \end{aligned} \quad (6.85)$$

the Reuss bounds reduce to $B_R = B_V$ and

$$G_R = \frac{5(c_{11} - c_{12})c_{44}}{4c_{44} + 3(c_{11} - c_{12})}. \tag{6.86}$$

For hexagonal crystals, having five independent elastic constants, the Voigt and Reuss bounds are

$$B_V = \frac{2(c_{11} + c_{12}) + 4c_{13} + c_{33}}{9}, \tag{6.87}$$

$$G_V = \frac{12c_{44} + 12c_{66} + c_s}{30}, \tag{6.88}$$

$$B_R = \frac{c^2}{c_s}, \tag{6.89}$$

and

$$G_R = \frac{5}{2} \frac{c_{44}c_{66}c^2}{(c_{44} + c_{66})c^2 + 3B_V c_{44}c_{66}}. \tag{6.90}$$

Here, c^2 and c_s are given by Equations (6.54) and (6.55). We point out that the polycrystalline bulk modulus B_R calculated assuming constant stress (Reuss) agrees with Equation (6.53) obtained for a hexagonal single crystal. On the other hand, the Voigt bulk modulus B_V calculated assuming constant strain agrees with Equation (6.58), obtained for a hexagonal single crystal with rigid axial ratio $(c/a)_0$. Obviously, when the hexagonal axial ratio does not depend on volume, $B_V = B_R = (2c_{13} + c_{33})/3$.

The G_V and G_R bounds can also be used to characterize the polycrystalline solids formed by randomly oriented anisotropic single crystal grains. In these quasi-isotropic materials it is useful to define a measure of elastic anisotropy as

$$A_{VR} = \frac{G_V - G_R}{G_V + G_R}. \tag{6.91}$$

This ratio has the following properties of practical importance: it is zero for isotropic crystals, for anisotropic crystals it is a single-valued measure of the elastic anisotropy, and it gives a relative magnitude of the actual elastic anisotropy. For most of the metals, $A_{VR} \leq 20\%$, but exceptionally large values could be observed for K, Na and Li [134].

The Hashin–Shtrikman Bounds

Hashin and Shtrikman [135] derived upper and lower bounds for B and G

using a variational method. Here we give results only for the cubic lattice. For lattices with lower symmetry the reader is referred to Grimvall [128] and references therein. For a cubic lattice with $c' < c_{44}$, these bounds are

$$\begin{aligned} B_u &= B_l = \frac{c_{11} + 2c_{12}}{3}, \\ G_l &= c' + 3\left(\frac{5}{c_{44} - c'} + 4\beta_1\right)^{-1}, \\ G_u &= c_{44} + 2\left(\frac{5}{c' - c_{44}} + 6\beta_2\right)^{-1}, \end{aligned} \tag{6.92}$$

where

$$\begin{aligned} \beta_1 &= \frac{3(B + 2c')}{5c'(3B + 4c')}, \\ \beta_2 &= \frac{3(B + 2c_{44})}{5c_{44}(3B + 4c_{44})}. \end{aligned} \tag{6.93}$$

For $c' > c_{44}$, the upper and lower bounds are reversed.

The Hershey Average

The Hershey's averaging method [54] turned out to give the most accurate relation between cubic single-crystal and polycrystalline data in the case of FeCrNi alloys [55]. According to this approach, the average shear modulus G is a solution of equation [54]

$$G^3 + \alpha G^2 + \beta G + \gamma = 0, \tag{6.94}$$

where

$$\begin{aligned} \alpha &= \frac{1}{8}(5c_{11} + 4c_{12}), \\ \beta &= -\frac{1}{8}c_{44}(7c_{11} - 4c_{12}), \end{aligned} \tag{6.95}$$

$$\gamma = -\frac{1}{8}c_{44}(c_{11} - c_{12})(c_{11} + 2c_{12}). \tag{6.96}$$

The Hill Average

Hill [136, 137] has shown that the Voigt and Reuss bounds are rigorous upper and lower bounds. The average bulk and shear moduli can be estimated from these bounds, *e.g.*, as

$$G_H = \frac{1}{2}(G_V + G_R)$$
$$B_H = \frac{1}{2}(B_V + B_R). \tag{6.97}$$

Alternatively, instead of the arithmetic average one might prefer to use the geometric or harmonic means. In weakly anisotropic materials, of course, all these averages lead to similar mean B and G.

6.4 Surface Energy and Stress

The surface energy is the surface excess free energy per unit area of a particular crystal facet and is one of the basic quantities in surface physics. It determines the equilibrium shape of mezoscopic crystals, it plays an important role in faceting, roughening, and crystal growth phenomena, and may be used to estimate surface segregation in binary alloys. With very few exceptions the available experimental surface energies stem from surface tension measurements in the liquid phase extrapolated to zero temperature [138, 139]. Because of the indirect measurement, the experimental surface energy data include large error bars ($20-50\%$). Furthermore, they correspond to isotropic crystals and thus yield no information about the crystal orientation dependence.

The theoretical surface free energy is calculated as half of the free energy needed to split an otherwise perfect monocrystal into two semi-infinite crystals. The surface energy per unit area is given by

$$\gamma = \frac{E^s}{A_{2\mathrm{D}}}, \tag{6.98}$$

where E^s is half of the free energy and $A_{2\mathrm{D}}$ is the area of the two-dimension (2D) unit cell. Since the surface energy is defined for a particular crystal facet, its magnitude might strongly depend on the orientation. Usually, the surface energy increases with the roughness of the surface. In terms of the cleavage force $F(z)$ acting between the two semi-infinite crystals separated by z, γ can be written as one half of the cleavage work per unit area,

$$\gamma = \frac{1}{2}\int_0^\infty F(z)dz. \tag{6.99}$$

The surface stress is the the reversible work per area to stretch the surface elastically. It has a decisive role for the understanding of a wide variety of surface phenomena, *e.g.* for surface reconstruction, shape transitions in nano-scale particles, surface alloying, surface diffusion, epitaxial growth, and self-assembled domain patterns. Experimental techniques have been used to establish the polar dependence of the surface stress [140], but a direct measurement of its magnitude is not feasible.

The stress tensor τ_{ij} (i, j stand for the in-plane coordinates x, y) is calculated from the change per unit area of the surface energy E^s upon a small deformation ϵ_{ij} of the surface unit cell, namely

$$\tau_{ij} = \frac{1}{A_{2D}} \frac{\partial E^s}{\partial \epsilon_{ij}}. \tag{6.100}$$

Using Equation (6.98), we can write

$$\tau_{ij} = \frac{1}{A_{2D}} \frac{\partial A_{2D}\gamma}{\partial \epsilon_{ij}} = \gamma\delta_{ij} + \frac{\partial \gamma}{\partial \epsilon_{ij}}, \tag{6.101}$$

where $\partial\gamma/\partial\epsilon_{ij}$ is the residual surface stress. This is the Shuttleworth equation connecting the surface energy and the surface stress [141].

It is apparent from the definitions that the surface stress and surface energy are of different nature. In the case of stable solids the free energy of a surface is always positive, otherwise the solid would gain energy by fragmentation. The surface stress, on the other hand, can either be positive or negative. We note that for a liquid, the surface free energy and the surface stress are equal due to the fact that in this case the surface energy does not change when the surface is strained, *i.e.* $\partial\gamma/\partial\epsilon_{ij} = 0$. These two quantities are frequently referred to by the common name "surface tension".

6.4.1 Numerical Calculation of the Surface Energy and Stress

A free surface is modeled by an atomic slab embedded in vacuum. The slab is formed by N atomic layers with specific Miller indices. The slab plus vacuum configuration is periodically repeated along the direction perpendicular to the atomic layers. The vacuum region is represented by a similar crystallographic lattice occupied by empty sites. The number of empty layers is N_v. The in-plane lattice constant is fixed to the bulk value, whereas the inter-layer distances near the surfaces might be relaxed to their equilibrium values to account for the surface relaxation. The typical slab and vacuum thicknesses for transition metals, needed to obtain bulk-like properties in the center of the slab and consequently realistic surfaces, are listed in Table 6.2.

Using the slab geometry, at zero temperature, the surface energy may be calculated as

$$\gamma = \frac{E_{\mathrm{slab}}(N) - NE_{\mathrm{bulk}}}{2A_{2D}}, \tag{6.102}$$

where E_{slab} is the total energy of the slab including the empty spheres simulating the vacuum, and E_{bulk} is the bulk energy per atom. The factor of 2 from the denominator arises from the two surfaces of the slab.

The surface stress tensor is calculated by "stretching" the two in-plane lattice vectors by ϵ, while the third lattice vector, which determines the

Table 6.2. The required number of atomic (N) and vacuum (N_v) layers used in a slab geometry to model the first few close-packed surfaces of transition metals. A_{2D} is the area of the 2D unit cell and a is the bulk lattice constant.

Structure	Surface	A_{2D}	N	N_v
fcc	(111)	$\sqrt{3}/4\,a^2$	8	4
	(100)	$1/2\,a^2$	8	4
	(110)	$\sqrt{2}/2\,a^2$	12	6
bcc	(110)	$\sqrt{2}/2\,a^2$	12	4
	(100)	a^2	12	6
	(211)	$\sqrt{3/2}\,a^2$	16	8
	(310)	$\sqrt{10}/2\,a^2$	16	8
	(111)	$\sqrt{3}\,a^2$	16	8
hcp	(0001)	$\sqrt{3}/2\,a^2$	8	4

layer−layer distance, is kept fixed. For this distortion the deformation tensor has the form

$$\epsilon_{ij} = \begin{bmatrix} \epsilon & 0 & 0 \\ 0 & \epsilon & 0 \\ 0 & 0 & 0 \end{bmatrix}. \tag{6.103}$$

Thus, the change in the surface energy $\delta E^s = \delta E_{\text{slab}}(N) - N\delta E_{\text{bulk}}$ is a function of ϵ. Calculations are carried out for both surface and bulk systems for several different ϵ values. Then, in order to minimize the numerical noise, we fit a polynomial to the calculated mesh-points,

$$\delta E_{\text{slab/bulk}} \approx c_0 + c_1\epsilon + c_2\epsilon^2 + ... \tag{6.104}$$

The surface stress τ is determined from the linear coefficients of the slab and bulk energies, *viz.*

$$\tau^{(s)} = \frac{c_{\text{slab},1} - c_{\text{bulk},1}}{4A_{2D}}, \tag{6.105}$$

where the factor of 4 arises from the two surfaces of the slab and the two elongated in-plane lattice vectors. Note that at the equilibrium volume, $c_{\text{bulk},1}$ vanishes and τ is determined merely by the linear coefficient of the slab energy. In the case of low symmetry surfaces, such as the *fcc* or *bcc* (110) surfaces, Equation (6.105) gives the average of the two main stress tensor components.

6.5 Stacking Fault Energy

A perfect *fcc* crystal has the ideal *ABCABCAB* stacking sequence, where the letters denote adjacent (111) atomic layers. The intrinsic stacking fault is the most commonly found fault in experiments on fcc metals. This fault is produced by a shearing operation described by the transformation $ABC \rightarrow BCA$ to the right hand side of an (111) atomic layer. It corresponds to the $ABCA\dot{C}\dot{A}\dot{B}\dot{C}$ stacking sequence, where the translated layers are marked by dots.[6] The formation energy of an extended stacking fault is defined as the excess free energy per unit area. At zero temperature, the stacking fault energy (SFE) is calculated as

$$\gamma_{\mathrm{SF}} = \frac{E_{\mathrm{fault}} - E_{\mathrm{bulk}}}{A_{\mathrm{2D}}}, \tag{6.106}$$

where E_{fault} and E_{bulk} are the energies of the system with and without the stacking fault, respectively, and A_{2D} denotes the area of the stacking fault.

6.5.1 Numerical Calculation of the Stacking Fault Energy

The intrinsic stacking fault creates a negligible stress near the fault core. Therefore, the faulted lattice approximately preserves the close-packing of the atoms, and can be modeled by an ideal close-packed lattice. Within the axial interaction model [142, 143] the stacking sequence along the $< 111 >$ direction is represented by a set of variables, S_i, where i is the layer index. The sign of S_i is determined by the translation connecting subsequent close-packed layers. Then the excess energy of a particular stacking sequence is expanded as $-\sum_i \sum_n J_n S_i S_{i+n}$, where the sums run on the atomic layers, and J_1, J_2, ... are the nearest-neighbor, next nearest-neighbor, *etc.*, interaction parameters. Using this representation, the energy of an intrinsic SFE can be expressed as

$$E_{\mathrm{fault}} - E_{\mathrm{bulk}} = -4J_1 - 4J_2 - 4J_3 - \mathcal{O}(J_4), \tag{6.107}$$

where $\mathcal{O}(J_4)$ stands for the contribution coming from the higher order terms $(J_4, J_5, ...)$. On the other hand, applying the axial interaction model to periodic structures, we obtain, for instance,

$$\begin{aligned} E_{fcc} &= J_0 + J_1 + J_2 + J_3 + \mathcal{O}(J_4) \\ E_{hcp} &= J_0 - J_1 + J_2 - J_3 + \mathcal{O}(J_4) \\ E_{dhcp} &= J_0 - J_2 - \mathcal{O}(J_4) \end{aligned} \tag{6.108}$$

where E_{fcc}, E_{hcp} and E_{dhcp} are the energies of fcc, hcp and $dhcp$ structures, respectively. By eliminating the interaction parameters from Equations (6.107) and (6.108), we can derive a relationship between the SFE and the energies of periodic structures. In the lowest order approximation, *i.e.* taking

[6] The extrinsic stacking fault corresponds to $ABCA\dot{C}\ddot{B}\ddot{C}\ddot{A}$ stacking, whereas the twin stacking fault is a plane boundary between two fcc stacking sequences with opposite orientation, *i.e.* $ABCABACBA$.

into account J_0 and J_1, the formation energy of the intrinsic stacking fault is twice the energy difference between the *hcp* and *fcc* phases, *viz.*

$$E^{2nd}_{\text{fault}} - E_{\text{bulk}} \approx 2(E_{hcp} - 3E_{fcc}). \tag{6.109}$$

Keeping terms up to the third order, for the stacking fault energy we find[7]

$$E^{3rd}_{\text{fault}} - E_{\text{bulk}} \approx E_{hcp} + 2E_{dhcp} - 3E_{fcc}. \tag{6.110}$$

This expression will be used in Section 9.3 to compute the stacking fault energy of *fcc* FeCrNi alloys.

6.6 Some Ideas about the Atomistic Modeling of the Mechanical Properties of Alloys

In general, by mechanical properties we understand the behavior of materials under external forces. These properties are of special importance in fabrication processes and applications. At the time of writing, a direct determination of the mechanical properties of complex alloys from *first-principles* theory is not feasible. On the other hand, well established phenomenological models exist, which are suitable for an accurate description of materials behavior under various mechanical loads. These models involve a large set of atomic-level physical parameters. Below, we give a very brief overview of some of these model.

Materials behavior are usually described in terms of stress or force per unit area and strain or displacement per unit distance. On the basis of stress and strain relations, one can distinguish elastic and plastic regimes. In elastic regime, at small stress, the displacement and applied force obey Hook's law and the specimen returns to its original shape on uploading. Beyond the elastic limit, upon strain release the material is left with a permanent shape. Several models of elastic and plastic phenomena in solids have been established. For a detailed discussion of these models, we refer to [144, 145, 146, 147, 148].

Within the elastic regime, the single crystal elastic constants (Section 6.2) and the polycrystalline elastic moduli (Section 6.3) play the principal role in describing the stress–strain relation. Within the plastic regime, the importance of lattice defects in influencing the mechanical behavior of crystalline solids was recognized long time ago. Plastic deformations are primarily facilitated by dislocation motion and can occur at stress levels far below those required for dislocation free crystals.

---

[7] Following the above procedure, the formation energy of the extrinsic stacking fault becomes $4(E_{dhcp} - E_{fcc})$ and that of the twin stacking fault $2(E_{dhcp} - E_{fcc})$.

The mechanical hardness represents the resistance of material to plastic deformation. It may be related to the yield stress separating the elastic and plastic regions, above which a substantial dislocation activity develops. In an ideal crystal dislocations can move easily because they experience only the weak periodic lattice potential. In real crystals, however, the motion of dislocations is impeded by obstacles, leading to an elevation of the yield strength. According to this scenario, the yield stress is decomposed into the Peierls and the solid-solution strengthening contribution. The stress needed to move a dislocation across the barriers of the oscillating crystal potential is the Peierls stress. In metals, this is found to be approximately proportional to the shear modulus G [148]. Therefore, the concentration dependence of the Peierls term is mainly governed by that of the elastic constants.

The solid-solution strengthening contribution is due to dislocation pinning by the randomly distributed solute atoms. Dislocation pinning by random obstacles has been described by classical models of Fleisher, Labusch and Nabarro [145, 146, 147] or more recent models by Clerc and Ledbetter [149]. According to the Labusch–Nabarro model, dislocation pinning is mostly determined by the size misfit (ε_b) and elastic misfit (ε_G) parameters. They are calculated from the concentration (x) dependent Burgers vector [144] $b(x)$ or lattice parameter, and shear modulus $G(x)$ as

$$\varepsilon_b = \frac{1}{b(x)}\frac{\partial b(x)}{\partial x}, \quad \text{and } \varepsilon_G = \frac{1}{G(x)}\frac{\partial G(x)}{\partial x}. \tag{6.111}$$

After lengthy calculations, it is obtained that the solid-solution strengthening contribution to the hardness depends on the concentration of solute atoms as $x^{2/3}{\varepsilon_L}^{4/3}$. Here, $\varepsilon_L \equiv \sqrt{{\varepsilon'_G}^2 + (\alpha\varepsilon_b)^2}$ is the Fleischer parameter, $\varepsilon'_G \equiv \varepsilon_G/(1+0.5|\varepsilon_G|)$ and $\alpha =9$–16.

Besides the above described bulk parameters, the formation energies of two-dimensional defects are also important in describing the mechanical characteristics of solids. The surface energy (Section 6.4) is a key parameter in brittle fracture. According to Griffith theory [148], the fracture stress is proportional to the square root of the surface energy, that is, the larger the surface energy is, the larger the load could be before the solid starts to break apart. Another important planar defect is the stacking fault in close-packed structures, such as the *fcc* and *hcp* lattices. In these structures, the dislocations may split into energetically more favorable partial dislocations having Burgers vectors smaller than a unit lattice translation [144]. The partial dislocations are bound together and move as a unit across the slip plane. In the ribbon connecting the partials, the original ideal stacking of close-packed lattice is faulted. The energy associated with this miss-packing is the stacking fault energy (Section 6.5). The balance between the SFE and the energy gain by splitting the dislocation determines the size of the stacking fault ribbon. The width of the stacking fault ribbon is of importance in many aspects of plasticity, as in the case of dislocation intersection or cross-slip. In both cases, the two partial dislocations have to be brought together to form an unextended dislocation before intersection or cross-slip can occur. By changing the SFE or the dislocation strain energy, wider or narrower dislocations can be produced and the mechanical properties can be altered accordingly. For instance, materials with high SFE permit dislocations to cross-slip easily. In materials with low SFE, cross-slip is difficult and dislocations are constrained to move in a more planar fashion. In this case, the constriction process becomes more difficult and hindered plastic deformation ensues. Designing for low SFE, in order to restrict dislocation movement and enhance hardness was adopted, *e.g.*, in transition metal carbides [150].

The principal problem related to modeling the mechanical properties of complex solid solutions is the lack of reliable experimental data of the alloying effects on the fundamental bulk and surface parameters. While the volume misfit parameters are available for almost all the solid solutions, experimental values of the elastic misfit parameters are scarce. There are experimental techniques to establish the polar dependence of the surface energy [140], but a direct measurement of its magnitude is not feasible [151, 152, 153]. In contrast to the surface energy, the stacking fault energy can be determined from experiments. For instance, one can find a large number of measurements on the stacking fault energy of austenitic stainless steels [154, 155]. However, different sets of experimental data published on similar steel compositions differ significantly, indicating large error bars in these measurements. On the other hand, during the last decade, the theoretical determination of these parameters in the case of metals and alloys has come within the reach of modern computational physics based on Density Functional Theory. In fact, a carefully performed computational modeling, based on modern *first-principles* alloy theory, can yield fundamental parameters with an accuracy comparable to or better than the experiments. In this sense, *ab intio* computer simulations may already have an important impact on the theoretical modeling of the mechanical properties of alloys.

6.7 Recommendations for EMTO Calculations

Any well designed data production with the EMTO method should be preceded by careful tests to assess the performance of the method for the actual problem. As a first step, one should always start by investigating the equation of state using different approximations for the exchange-correlation functional. The output of these studies allows one to

- establish the accuracy of the EMTO method, and
- decide which approximation is the most suitable

for the system in question. The first point is achieved by comparing the EMTO results with other *ab initio* results and the second one by comparing the theoretical results with the available experimental data. Generally, in the case of metals the highest accuracy is achieved using s, p, d, f basis set in combination with soft-core approximation[8] and a gradient corrected exchange-correlation functional. However, sometimes especially for systems with unit cells containing more than 30–50 atoms, a smaller basis set (only s, p, d orditals) is preferred in order to reduce the computation time. In such situations, one may opt to employ less accurate approximations (*e.g.*, LDA or frozen-core) which could compensate for the errors in the equilibrium volume and bulk modulus. For engineering purposes this level of "data tuning" is not only admissible but many times also indispensable.

The slope matrix from the EMTO kink-cancelation equation (Section 2.1.3) and overlap matrix (Section 2.1.4) is obtained by solving the Dyson equation (Section 3.1) on a real space cluster (Section 3.3). As a result, there always will be some errors in the total energy which are connected with the finite size of the real space cluster. Because of this, in EMTO calculations involving lattice distortions, the real space cluster size used for the slope matrix should be kept constant upon lattice distortion. This is the case, for instance, in optimization of the shape of the unit cell, elastic constant and stacking fault calculations, *etc.*

The slope matrix parametrization (Section 3.4) depends on the actual electronic structure and volume. In order to decide about the type of the energy expansion, one should investigate the width of the valance band including the semi-core states relative to the muffin-tin zero. In most cases, a simple Taylor expansion around $\omega_0 = 0$ (Section 3.4.1) ensures a sufficiently high accuracy at zero pressure. However, for systems with high-lying semi-core states a second expansion around ω_1 (typically between -40 and -20) should be included (Section 3.4.3). In fact, all high-pressure studies should be carried out with semi-core states treated as valance states and using the two-center Taylor expansion for the slope matrix.

In the self-consistent calculations, as long as the overlap error remains sufficiently small (Section 2.3.1), the radius of the potential sphere (s_R) should be fixed to the radius of the atomic sphere (w_R). Within the Spherical Cell Approximation, using both larger and smaller potential spheres on different sites (*i.e.*, $s_R > w_R$ for some sites and $s_{R'} < w_{R'}$ for some other sites) might lead to the divergence of the optimized overlapping muffin-tin potential (Section 2.1.1).

When choosing the energy contour in Equations (2.45) and (2.49) (Sections 2.1.5 and 2.2) one should make sure that the contour is large enough to include the valence band below the actual Fermi level, but, at the same time, it does not include any high-lying core states. When semi-core states are present, they should also be included in the contour. For a stable calculation, it is advisable to set the contour in such a way that the gap between the bottom

[8] Within the soft-core approximation the core states are recalculated after each iteration. Opposite to this, the frozen-core approximation uses fixed atomic-like core states.

of the contour and the valence band should be around 0.1–0.5 Ry. A similar gap should also be seen between the bottom of the contour and the core states with the highest energy. This is especially important if the contour integration is performed using a numerical technique (*e.g.*, the Gaussian quadrature) that does not include information from the endpoints and therefore the actual integration limit along the real energy axis depends on the imaginary part of the last point (the complex energy with the lowest real component) from the contour.

Calculations for random systems should start with the determination of the screening constant from the screened impurity model (Sections 5.2 and 8.1). Although many properties (*e.g.*, elastic constants) show negligible dependence on this parameter, an inappropriate value can lead to large errors in the volume and therefore also in the computed quantities.

The Brillouin zone sampling requires special attention in dilute systems. The k-mesh, which is "sufficient" for the end-members and for solid solutions containing more than ∼10–15% solvent, can very well give results that scatter at low concentrations.

The accuracy of the shape function technique for a given lattice should be investigated (Section 4.1). Performing test calculations for model systems with uniform charge distribution allows one to check both the convergence of the radial mesh and the l-summations for the shape function (Section 4.2.6).

Finally, the charge neutrality requirement (Section 4.2.6) should always be monitored. It represents the most severe test for the accuracy of the (i) one-center expansion for the charge density (Sections 2.2 and 5.2.2), (ii) shape function (Section 4.1) and (iii) slope matrix (Section 3.4). With large charge misfit no accurate total energies and derived quantities will be obtained.

7

Ordered Solids

This chapter is dedicated, first of all, to demonstrating the application of the EMTO method in the case of ordered metallic and non-metallic systems. In Section 7.1, we start by investigating the equation of state of a large set of simple and transition metals crystallizing in body centered cubic, face centered cubic or hexagonal close-packed crystal structure. Results for the equilibrium atomic radius and bulk modulus will be presented for three different exchange-correlation energy functionals. The EMTO determination of the formation energy of mono-vacancies will be illustrated in the case of *fcc* Al. We shall also discuss some surface energy and surface stress results obtained for low-index surfaces of 4*d* transition metals. The ground state properties of some common non-metallic solids and two silicate perovskites will be explored in Section 7.2. The elastic constant calculation will be exemplified in the case of solid helium. Some results obtained for transition-metal nitride surfaces will be discussed in Section 7.3. The most significant numerical details of the EMTO calculations presented in this chapter are listed in Section C.1.[1]

7.1 Simple and Transition Metals

7.1.1 Equilibrium Bulk Properties

Many physical properties depend sensitively on volume. Because of this, a modern *ab initio* total energy method should, first of all, be able to reproduce the equation of state and related quantities (equilibrium volume, bulk modulus and Grüneisen constant) with high accuracy. The performance of the EMTO method for the equilibrium atomic volume and bulk modulus is demonstrated in Figures 7.1 and 7.2. Here, we compare the EMTO values with those obtained using the full-potential linear muffin-tin orbital (FP-LMTO)

[1] For details about the FP calculations quoted in this chapter, the reader is referred to the corresponding references.

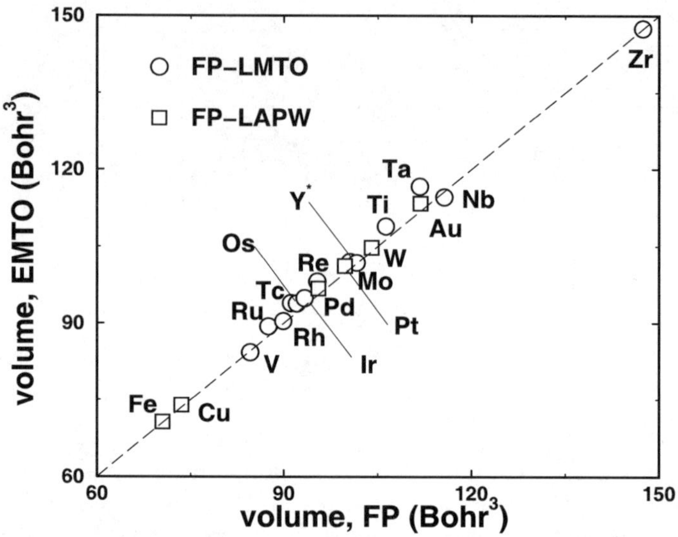

Fig. 7.1. Comparison between different theoretical equilibrium atomic volumes (in Bohr^3) for selected transition metals calculated using the EMTO method (ordinate) and two full-potential methods (abscissa) [156, 157]. The volume for Y has been divided by a factor of 2 in order to fit on the present scale. Circles denote the FP-LMTO results [156], and squares the FP-LAPW results [157].

[156] and linear augmented plane wave (FP-LAPW) [157] methods. For this test, we selected a few $3d$, $4d$ and $5d$ transition metals, for which systematic FP-LMTO and FP-LAPW studies are available. All theoretical data from Figures 7.1 and 7.2 were computed using the *Local Density Approximation* (LDA) [6, 7] for the energy functional. The Perdew and Wang parametrization of the LDA functional is given in Appendix A. All calculations were carried out for the space groups corresponding to the low temperature experimental crystal structures [129]. The ferromagnetic order was taken into account in the case of *bcc* Fe, and all the other elements were treated as nonmagnetic solids.

The mean deviations between the EMTO and the FP atomic volumes from Figure 7.1 are 0.33%, 0.43%, and 0.49% for the $3d, 4d$, and $5d$ metals, respectively. The same figures for the bulk moduli[2] are 0.20%, 1.20% and 3.28%, respectively. The excellent agreement between the EMTO and FP data shows that for equations of state the EMTO method has the accuracy of formally exact full-potential methods.

In the above tests, for the hexagonal metals (Ti, Y, Zr, Tc, Ru, Re, Os) we used the theoretical axial ratios (c/a) calculated within the LDA. The equilibrium hexagonal axial ratio is determined by minimizing the total energy

---

[2] The standard unit for the bulk modulus is GPa or MPa: 1 Ry/Bohr^3 $\approx$ 14710.51 GPa.

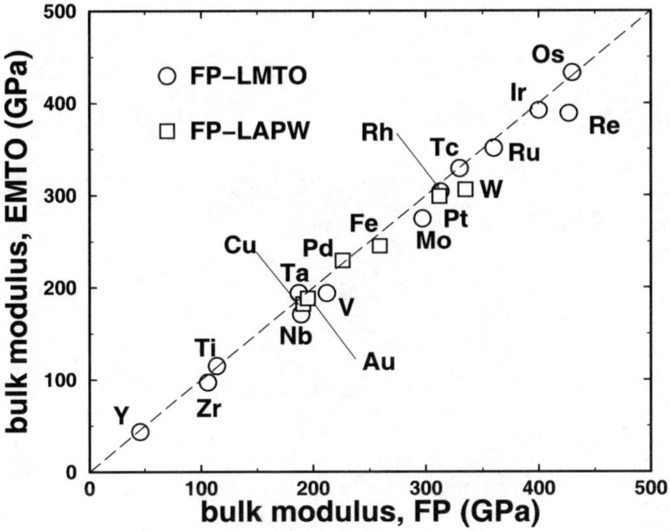

Fig. 7.2. Comparison between different theoretical equilibrium bulk moduli (in GPa) for selected transition metals calculated using the EMTO method (ordinate) and two full-potential methods (abscissa) [156, 157]. For notation see caption for Figure 7.1.

with respect to the volume and c/a. This type of calculation necessitates increased accuracy compared to that needed for the equation of state. This is because the total energy has to be computed against anisotropic lattice distortion, which usually results in a much smaller energy change than that due to the isotropic distortion, *i.e.* volume change.

In Figure 7.3, we compare the EMTO c/a values for 13 *hcp* metals with those calculated using the FP-LMTO method [156] and with the available experimental data [158]. At temperatures below 673 K, lanthanum has a double-*hcp* (*dhcp*) structure with c/a ratio of 3.225 [158, 159]. In Figure 7.3, the EMTO result for La was obtained for the *hcp* phase, and therefore we compare this with half of the experimental *dhcp* axial ratio. Both sets of theoretical results from Figure 7.3 were obtained using the *Generalized Gradient Approximation* (GGA) for the exchange-correlation functional [8, 9, 15, 16]. Note that the FP-LMTO calculation used the functional proposed by Perdew and Wang [8], whereas the EMTO calculations employed the functional by Perdew, Burke, and Ernzerhof [15].[3] Expressions for the latter are given in Appendix A.

Except for Y and Ti, the agreement between the EMTO and the FP-LMTO hexagonal axial ratios is very good. Yttrium and titanium have high-

[3] In literature, these two GGA level approximations are often referred to as the PW91 [8] and PBE [15] approximations.

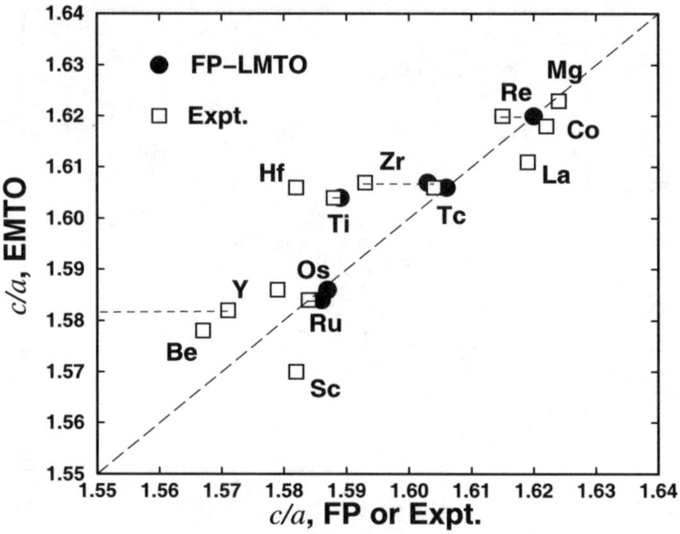

Fig. 7.3. Comparison between the theoretical and experimental equilibrium hexagonal lattice parameter (c/a) for hexagonal simple and transition metals. Theoretical values were calculated using the EMTO method (ordinate) and a full-potential method [156] (abscissa). The experimental data for Sc is from [159], and the rest from [158] (abscissa). The horizontal dashed lines mark the deviations between the FP-LMTO and experimental values. Note that the FP-LMTO result for Y (1.529) is not shown.

lying semi-core states, which have a significant influence on the equilibrium properties. Furthermore, in these metals the total energy versus c/a is very shallow near the energy minimum. Accordingly, we ascribe the discrepancies between the theoretical data for Y and Ti to the numerical difficulties associated with such calculations. The agreement between theory and experiment is also satisfactory. Somewhat larger deviations (0.7−1.5%) can be seen for Hf, Ti and Sc. However, we should point out that for these elements the measured c/a values also show some scatter. For instance, two independent experimental measurements reported 1.58 [159] and 1.60 [160] for the equilibrium hexagonal axial ratio of Sc.

The EMTO equilibrium atomic radii and bulk moduli for 38 metals are listed in Table 7.1. Except for Li and La, all the calculations were carried out for the space groups corresponding to the low temperature experimental crystal structures [129]. At low temperature, Li has a close-packed samarium-type hexagonal structure [129]. We approximated this structure by the fcc phase. For La we used the *hcp* structure instead of the low-temperature *dhcp* structure. However, due to the similar packing fractions, the above approximations are expected to have a minor effect on the calculated equations of state. In Table 7.1, in addition to the LDA and GGA data, we also list results obtained

Table 7.1. Theoretical (EMTO) and experimental [129] equilibrium atomic radii (w in Bohr) and bulk moduli (B in GPa) for metals with close-packed crystal structures from periods 2, 3, 4, 5 and 6. Periods are separated with horizontal lines. The EMTO results were obtained using the LDA [7], GGA [15] and LAG [17] energy functionals. The corresponding mean absolute values of the relative deviations from the experimental values are listed in Table 7.2.

	Z	str.	w_{LDA}	w_{GGA}	w_{LAG}	$w_{\text{expt.}}$	B_{LDA}	B_{GGA}	B_{LAG}	$B_{\text{expt.}}$
Li	3	fcc*	3.13	3.18	3.21	3.237	15.0	14.0	13.3	12.6
Be	4	hcp	2.30	2.34	2.33	2.350	132	123	123	110
Na	11	bcc	3.77	3.91	3.92	3.928	8.56	7.88	7.54	7.34
Mg	12	hcp	3.27	3.34	3.33	3.346	38.5	36.8	36.7	34.1
Al	13	fcc	2.95	2.99	2.98	2.991	85.5	76.8	78.3	72.8
K	19	bcc	4.69	4.92	4.92	4.871	4.00	4.16	3.95	3.70
Ca	20	fcc	3.95	4.10	4.06	4.109	17.6	16.5	16.3	18.4
Sc	21	hcp	3.28	3.36	3.33	3.424	66.6	69.6	67.4	54.6
Ti	22	hcp	2.96	3.04	3.01	3.053	115	110	111	106
V	23	bcc	2.72	2.79	2.76	2.813	194	177	183	155
Cr	24	sc	2.60	2.65	2.62	2.684	285	259	268	160
Fe	26	bcc	2.56	2.64	2.60	2.667	245	191	213	163
Co	27	hcp	2.53	2.60	2.56	2.613	285	197	234	186
Ni	28	fcc	2.53	2.60	2.57	2.602	244	196	214	179
Cu	29	fcc	2.60	2.69	2.65	2.669	182	142	155	133
Rb	37	bcc	5.00	5.27	5.26	5.200	3.19	3.43	3.24	2.92
Sr	38	fcc	4.30	4.45	4.41	4.470	14.0	13.1	13.1	12.4
Y	39	hcp	3.65	3.76	3.72	3.760	43.6	46.2	44.5	41.0
Zr	40	hcp	3.28	3.36	3.32	3.347	97.1	87.6	90.0	94.9
Nb	41	bcc	3.01	3.08	3.05	3.071	171	142	152	169
Mo	42	bcc	2.90	2.94	2.92	2.928	275	250	259	261
Tc	43	hcp	2.82	2.86	2.84	2.847	329	289	306	297[a]
Ru	44	hcp	2.77	2.82	2.80	2.796	351	300	322	303
Rh	45	fcc	2.78	2.84	2.81	2.803	304	251	272	282
Pd	46	fcc	2.85	2.92	2.89	2.840	229	166	191	189
Ag	47	fcc	2.97	3.07	3.03	3.018	137	88.7	104	98.8
Cs	55	bcc	5.36	5.73	5.72	5.622	2.08	2.32	2.16	2.10
Ba	56	bcc	4.38	4.67	4.59	4.659	8.28	7.75	7.57	9.30
La	57	hcp*	3.79	3.95	3.90	3.925	28.2	29.8	28.8	26.6
Hf	72	hcp	3.20	3.27	3.23	3.301	142	134	137	108
Ta	73	bcc	3.03	3.10	3.06	3.073	194	179	183	191
W	74	bcc	2.92	2.97	2.95	2.937	306	294	298	308
Re	75	hcp	2.86	2.90	2.88	2.872	389	355	368	360
Os	76	hcp	2.82	2.86	2.84	2.824	433	387	406	418[a]
Ir	77	fcc	2.83	2.87	2.85	2.835	392	340	362	358
Pt	78	fcc	2.89	2.95	2.92	2.897	299	245	266	277
Au	79	fcc	3.00	3.08	3.05	3.012	188	136	155	166
Pb	82	fcc	3.60	3.71	3.67	3.656	59.6	36.9	43.7	41.7

[a]Estimated values, [133]. *Not ground state structure.

Table 7.2. The mean absolute values of the relative deviations (in percent) between the EMTO and experimental atomic volumes and bulk moduli from Table 7.1. Errors are given for each period i separately (Δ_i) and also for all 38 metals (Δ_{tot}).

	w_{LDA}	w_{GGA}	w_{LAG}	B_{LDA}	B_{GGA}	B_{LAG}
Δ_2	2.72%	1.09%	0.84%	19.5%	11.5%	8.7%
Δ_3	2.55%	0.22%	0.35%	15.7%	6.9%	6.0%
Δ_4	3.38%	0.80%	1.71%	32.3%	16.9%	22.5%
Δ_5	1.83%	0.91%	0.74%	12.0%	9.2%	5.5%
Δ_6	1.84%	1.23%	0.89%	11.4%	10.8%	7.1%
Δ_{tot}	2.34%	0.94%	1.02%	17.8%	11.7%	10.7%

using the *Local Airy Gas Approximation* (LAG) for the exchange-correlation functional [17, 18]. The LAG approximation is introduced in Appendix A.

The mean absolute values of the relative deviations between the EMTO and experimental atomic volumes and bulk moduli listed in Table 7.1 are given in Table 7.2. The LDA atomic radii deviate, on average, by 2.34% from the experimental values, while those calculated in the GGA and the LAG model deviate by 0.94% and 1.02%, respectively. The corresponding LDA, GGA and LAG errors for the bulk modulus are 17.8%, 11.7% and 10.7%, respectively. Hence, among the energy functionals considered in Table 7.1, the LDA is found to give the largest mean deviations (Δ_{tot}), whereas the GGA and LAG lead to similar small mean deviations for both atomic radii and bulk moduli. On the other hand, considering the mean errors for the energy functionals separately for periods 2, 3, 4, 5 and 6, the relative merits become quite different. It is clear that the simple LAG model performs the best for Li, Rb and Cs periods as it gives the smallest Δ_2, Δ_5 and Δ_6 values for both atomic radii and bulk moduli. For the third period, the GGA and LAG approximations have similar accuracies. However, when only the fourth period (especially the $3d$ series) is taken into account, the accuracy of the GGA is superior to that of the LAG and LDA approximations. It is interesting to note that with increasing period the accuracy of the LDA approximation increases and for the sixth period it gives errors which are relatively close to those obtained within the GGA.

7.1.2 Vacancy Formation Energy

The theoretical description of vacancies has been used several times as a benchmark of total energy calculation methods [161, 162, 163]. The formation energies of mono vacancies in *fcc* Al are listed in Table 7.3. There are two important issues that should be considered in this type of calculation. First, the size of the supercell should be sufficiently large in order to minimize the vacancy−vacancy interaction. Second, in alkali, alkaline earth and p metals the valence electrons are free-electron like, and therefore the Brillouin zone sampling should be performed with high accuracy. The latter problem is less

Table 7.3. Theoretical (LDA) and experimental formation energies (E_v in eV) and relaxation of the first nearest-neighbor lattice sites (δ_{NN} in %) for a vacancy in *fcc* Al. The EMTO calculations have been performed on 16 (numbers in parenthesis) and 32 atom supercells.

	EMTO	Full-potential	Experimental
E_v, un-relaxed	0.84(0.89)	0.82^a,0.73^b	-
E_v, relaxed	0.66(0.64)	0.66^a,0.62^b,0.66^c	0.67 ± 0.03^d
δ_{NN}	-1.41(-1.58)	-1.50^b	-

[a] pseudopotential [161]
[b] KKR [163]
[c] pseudopotential [162]
[d] experimental [164]

severe for transition metals in which the d states are more localized in contrast to the long-ranging s and p states.

The EMTO results from Table 7.3 were obtained for *fcc* supercells of 16 and 32 atoms using the LDA for the exchange-correlation functional.[4] It has been shown [79] that increasing the size of the supercell beyond 32 atoms, as long as the proper convergence of the Brillouin zone sampling is ensured,

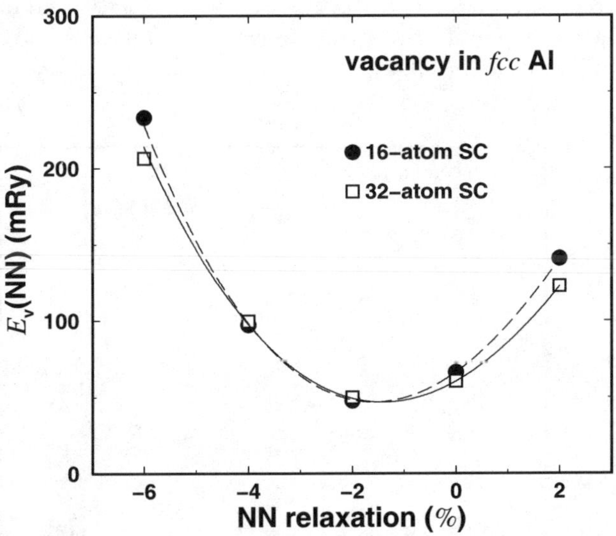

Fig. 7.4. Formation energy as a function of the nearest-neighbor distance (NN) for a vacancy in *fcc* Al. The energy minima correspond to −1.58% and −1.41% inward relaxation of the first coordination shell around the vacancy in 16 (dashed line) and 32 (solid line) atoms supercells, respectively.

[4] Usually, the vacancy formation energies are expressed in eV per vacancy: 1 Ry ≈ 13.605698 eV.

has no significant effect on the formation energy. The un-relaxed energies from Table 7.3 were obtained for ideal *fcc* based supercells, while the relaxed values correspond to the radial relaxation of the nearest-neighbor sites (NN) of the vacancy. The changes of the vacancy formation energies with the NN distances are shown in Figure 7.4.

The un-relaxed and relaxed EMTO energies for 32 atoms supercell agree very well with the pseudopotential results from [161, 162], but they are 13% and 6%, respectively, larger than those obtained using the full-potential KKR method [163]. This difference may be assigned to the fact that in Ref. [163] the relaxation of the effective potential beyond the nearest-neighbor sites around the vacancy was neglected. The EMTO method gives a small inward relaxation for the NN lattice sites, in good accordance with the full-potential KKR results [163].

7.1.3 Surface Energy and Stress of Transition Metals

The calculation of the surface stress for an ideal unrelaxed surface is illustrated in Figure 7.5. Here we plotted the total energy for the (111) surface of *fcc* Pt as a function of the deformation parameter ϵ introduced in Equation (6.103). The energies were calculated using the EMTO method and the LDA for the exchange-correlation term [165]. Results are shown for two atomic radii: one corresponding to the equilibrium volume (w_0) and another corresponding to a slightly smaller volume ($0.99w_0$).

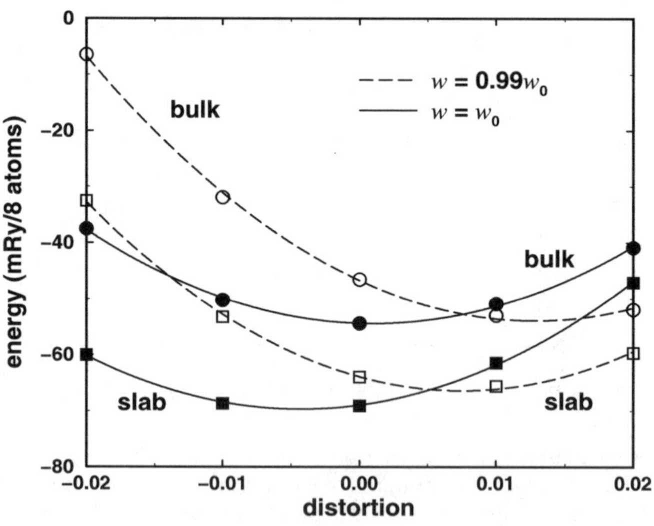

Fig. 7.5. Total slab and bulk energies for Pt as a function of the distortion parameter defined in Equation (6.103) for an ideal *fcc*(111) surface. Energies are shown for the equilibrium radius w_0 and $0.99w_0$.

Using Figure 7.5, one can determine the surface stress for *fcc* Pt and trace the effect of volume change on the surface stress. We can see that at the equilibrium volume the bulk total energy curve has a minimum at $\epsilon = 0$, *i.e.* $c_{\text{bulk},1} = 0$, while the slab energy has a minimum at $\epsilon < 0$, corresponding to a slope of $c_{\text{surf},1} > 0$ (see Section 6.4 for notations). It is interesting that for $w < w_0$ the bulk and slab minima are shifted, but the difference between the slopes of the energy curves, that enters in Equation (6.105), is not significantly altered. Since $(c_{\text{slab},1} - c_{\text{bulk},1}) > 0$, the (111) surface of *fcc* Pt exhibits a tensile (positive) surface stress. The positive surface stress is in line with the simple model described by Ibach [166]. According to that, upon cleaving a metal surface the electronic charge density of the broken bonds is redistributed between surface atoms and their back-bonds. This leads to increased charge density between surface atoms and leads to tensile surface stress on metal surfaces.

In contrast to the surface energy [171], the surface stress is drastically influenced by surface geometry [168], and therefore for an accurate surface stress calculation one should first determine the surface atomic structure. Real surfaces of pure metals can adopt different structures from those of ideally truncated crystals. During relaxation only rigid inward or outward displacement of the atomic layers occurs, while in the case of reconstruction, the displacement

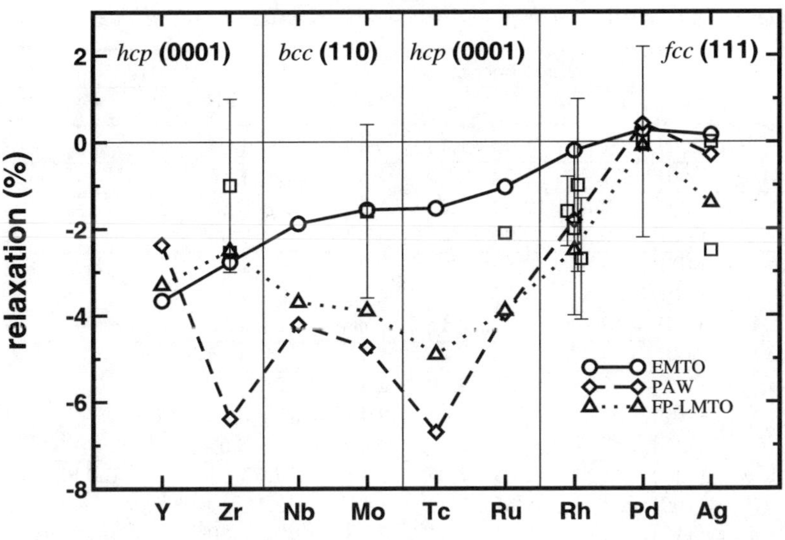

Fig. 7.6. Surface relaxation for the close-packed surfaces of 4*d* transition metals. Solid line stands for the EMTO results [167], dashed line shows full-potential relaxations [168] calculated using the projected augmented wave (PAW) method [169], and dotted line belongs to FP-LMTO results [170]. Open squares with error bars display experimental results [167]. (To avoid overlapping, the symbols and error bars of experimental values for Rh(111) are slightly shifted along the abscissa.)

of atoms may alter the two-dimensional symmetry of the surface. Experimental studies have demonstrated that the surface layer of clean transition metal surfaces relaxes inward [172], *i.e.* the inter-layer distance between the topmost two atomic layers is smaller than that in the bulk. Expansion of the top layer has been found for some surfaces of noble metals. The top layer relaxation is often accompanied by relaxation of the subsurface atomic layers, resulting in an oscillatory multi-layer relaxation.

Since no reliable atomic forces can be computed using the EMTO method, the multi-layer relaxation cannot be addressed by this approach. Nevertheless, the top layer relaxation can be determined from the EMTO total energies calculated as a function of the distance between the first and second atomic layers denoted by λ_{12}^s. The equilibrium top layer relaxation corresponds to the minimum of the surface energy as a function of $\delta \equiv (\lambda_{12}^s - \lambda_{12}^b)/\lambda_{12}^b$, where λ_{12}^b is the bulk equilibrium inter-layer distance.

The EMTO-GGA surface relaxations obtained for the close-packed surfaces of the 4*d* transition metals are compared to the full-potential results [168, 170] in Figure 7.6. The full-potential values were calculated using the projected augmented wave (PAW) method [169] and the FP-LMTO method. For comparison, some experimental values are also indicated [167]. Except for

Table 7.4. Surface energy and surface stress for the low index surfaces of 4*d* transition metals. The surface energy and stress are in units of J/m^2.

	str.	surf.	Surface energy EMTO	FP[a]	Expt.[b]	Surface stress EMTO	FP	Expt.
Y	hcp	(0001)	1.09	1.00	1.13	0.60	1.00[a]	
Zr	hcp	(0001)	1.76	1.57	2.00	1.57	1.15[a]	
Nb	bcc	(100)	2.58	2.32	2.70	1.79	0.89[a]	
		(110)	2.31	2.06		3.44	2.99[a]	
Mo	bcc	(100)	3.93	3.15	3.00	3.27	1.98[a]	
		(110)	3.40	2.73		4.15	2.96[a]	
Tc	hcp	(0001)	3.14	2.21	3.15	3.48	2.59[a]	
Ru	hcp	(0001)	3.53	2.52	3.05	4.75	3.15[a]	
Rh	fcc	(111)	2.71	2.01	2.70	2.11	2.73[a], 2.97[c]	
		(100)	2.84	2.35		2.14	2.35[a], 3.15[c], 3.66[d]	
		(110)	3.20			0.93	2.54[c]	
Pd	fcc	(111)	1.64	1.33	2.05	2.15	2.57[a], 3.69[e]	6.0±0.9[f]
		(100)	1.74	1.51		1.69	2.16[a], 1.98[d]	
		(110)	2.05			0.68	2.30[d]	
Ag	fcc	(111)	0.96	0.76	1.25	0.75	0.79[a]	1.41[g]
		(100)	0.91	0.84		0.88	1.31[a], 1.66[d]	
		(110)	1.10			0.43		

[a] PAW, [168] [d] FP-LMTO, [175] [g] Expt., [178]
[b] Expt., [138] [e] LCAO, [176]
[c] Pseudopotential, [174] [f] Expt., [177]

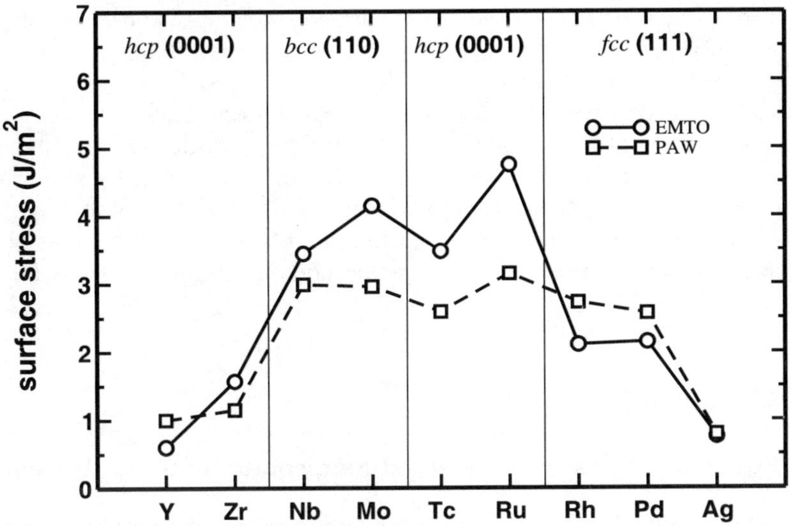

Fig. 7.7. Theoretical surface stress for the close-packed surfaces of 4*d* transition metal. Solid and dashed lines represent EMTO and PAW [168] results, respectively.

Pd(111) and Ag(111) surfaces, the directions of the EMTO surface relaxations are the same as those reported in other theoretical studies. For Pd and Ag, EMTO gives a small positive relaxation, which is in agreement with other predictions [173]. However, in most cases, the magnitude of relaxation is smaller in EMTO calculations than in other theoretical reports. When comparing the EMTO results with other theoretical data, one should take into account that in the PAW calculations [168] the subsurface layers were also relaxed, whereas in the FP-LMTO calculations [170] the *hcp*(0001) surfaces were modeled by *fcc*(111) surfaces. We also note that the overlapping muffin-tin approximation and the spherical cell approximation (Section 2.3), using a single muffin-tin zero everywhere in the supercell, may also be a possible source of inaccuracy in EMTO calculations for surface relaxations.

Figure 7.6 also demonstrates that (i) there are considerable differences between theoretical and experimental surface relaxations, (ii) experimental relaxations show significant dispersion for certain elements, and (iii) there are significant deviations between relaxations derived in different *ab initio* calculations. The EMTO calculations give a decreasing inward relaxation across the series (and slight outward relaxation for Pd and Ag), the PAW data exhibit two maxima (for Zr and Tc), and the FP-LMTO values (except for Y and Ag) follow approximately a parabolic trend. The EMTO trend has been explained in terms of the reduction of the number of *sp* electrons in the surface layer relative to bulk [167].

In Table 7.4, we list the theoretical surface stress[5] calculated for the relaxed geometries of the 4*d* metals together with the available experimental data. In agreement with Ibach [166], the surface stress is positive for each metals. The surface stress decreases with increasing roughness, except for Rh and Ag, where the (100) surfaces have slightly larger surface stress than the (111) surfaces. In Figure 7.7, we compare the EMTO surface stress values obtained for the most close-packed surfaces with those calculated using the PAW method [168]. It is surprising that in spite of the large deviations in the surface relaxation (Figure 7.6), the two methods give similar trends for the surface stress.

7.2 Non-metallic Solids

7.2.1 Equation of State for Selected Semiconductors and Insulators

The EMTO equilibrium atomic radii and bulk moduli for four common non-metallic systems are compared with the experimental data in Table 7.5. The theoretical values are listed for the LDA and GGA and LAG approximations for the exchange-correlation functional. We observe that LDA and GGA give similar mean absolute values of the relative errors. Like for the 5*d* transition metals, the LDA over-binding is overcompensated by the gradient correction within the GGA. On the other hand, for solids from Table 7.5, the simple LAG gradient level approximation outperforms both LDA and GGA, giving 0.8% and 9.1% for the mean errors in atomic radius and bulk modulus, respectively.

Table 7.5. Theoretical and experimental [157] equilibrium atomic radii and bulk moduli for some selected non-metallic solids. The theoretical results were obtained using the EMTO method in combination with LDA [7], GGA [15] and LAG [17] energy functionals. The mean absolute values of the relative deviations between the EMTO values and the experimental data are given (Δ_{tot}). The equilibrium atomic radii (w) are expressed in Bohr and the bulk moduli (B) in GPa. The crystal structures designations are: A4 diamond, B3 zincblende, and B1 NaCl type.

	str.	w_{LDA}	w_{GGA}	w_{LAG}	$w_{expt.}$	B_{LDA}	B_{GGA}	B_{LAG}	$B_{expt.}$
Si	A4	3.163	3.198	3.189	3.182	100	92.8	94.0	98.8
Ge	A4	3.303	3.384	3.354	3.318	71.6	61.2	64.0	76.8
GaAs	B3	3.296	3.375	3.346	3.312	73.0	62.0	72.1	74.8
NaCl	B1	3.202	3.346	3.337	3.306	32.9	23.0	21.7	24.5
Δ_{tot}		1.2%	1.4%	0.8%		11.2%	12.4%	9.1%	

[5] Here, the surface energy and stress are expressed in units of J/m^2: 1 $Ry/Bohr^2 \approx$ 778.446 J/m^2.

7.2.2 Elastic Properties of Solid Helium under Pressure

With the refinement of high-pressure techniques, there has been accelerated interest in the properties of rare gas solids. An obvious question that is raised in connection with high-pressure measurements is the hydrostatic limit of the pressure medium [179]. Below this pressure limit, the medium acts as a quasi-hydrostatic pressure-transmitting environment. At higher pressures nonhydrostaticity develops, which might affect the measured physical properties. Solid helium is one of the most important among the rare gas solids, because it is considered as the best quasi-hydrostatic medium [179, 180, 181, 182].

During the last few decades, numerous experimental investigations focused on the high-pressure physical properties of solid He. Zha *et al.* [183] measured the complete set of room-temperature single-crystal elastic constants between 13 and 32 GPa, whereas Nabi *et al.* [184] reported a density functional description of the equation of state and elastic properties up to 150 GPa.

At low temperature and pressures below $\sim$ 12 GPa, He crystalizes in *hcp* structure [185, 186, 187]. With increasing temperature a fcc phase is stabilized from $\sim$ 15 K and $\sim$ 0.1 GPa to $\sim$ 285 K and $\sim$ 12 GPa. Apart from this small fcc stability field, the *hcp* structure remains the stable phase of solid He up to 58 GPa, the highest pressure considered in experiments [185]. The stability of *hcp* phase is fully supported by the EMTO calculations carried out using

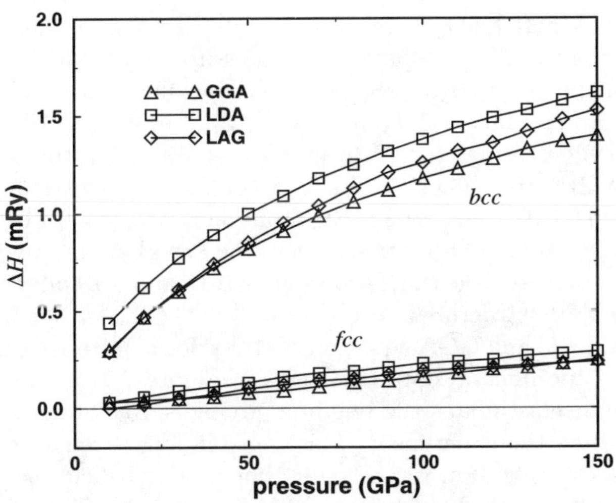

Fig. 7.8. Theoretical enthalpy for the *bcc* and fcc structures relative to the stable *hcp* phase as a function of pressure. Calculations were carried out using the EMTO method in combination with the LDA (squares), LAG approximation (diamond) and GGA (triangle) for the exchange-correlation functional.

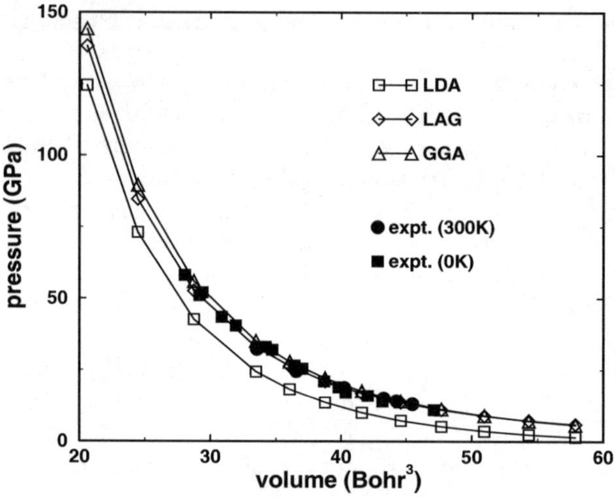

Fig. 7.9. Theoretical and experimental equation of state for solid He. The theoretical pressure–volume curves have been obtained using the EMTO method in combination with the LDA (open squares), LAG approximation (open diamond) and GGA (open triangle) for the exchange-correlation functional. Filled circles: room-temperature data [183], filled squares: temperature reduced data [185].

the LDA, GGA and LAG energy functionals.[6] In Figure 7.8, we show the computed enthalpy (H) as a function of pressure and crystal structure. The enthalpies for the fcc (ΔH_{fcc}) and bcc (ΔH_{bcc}) structures are plotted relative to that of the hcp phase ($\Delta H_{fcc/bcc} \equiv H_{fcc/bcc} - H_{hcp}$). All three density functionals predict the hcp structure as the most stable low-temperature phase of He. At 10 GPa the calculated LDA, LAG and GGA enthalpy differences are $\Delta H_{fcc} = 0.03, 0.01$ and 0.02 mRy/atom and $\Delta H_{bcc} = 0.44, 0.28$ and 0.29 mRy/atom. The pressure further stabilizes the hcp phase relative to the cubic structures, with an average $\partial \Delta H/\partial p$ slope of 0.15 and 0.70 mRy per 100 GPa, for the fcc and bcc structures, respectively.

The LDA, LAG and GGA equation of states for hcp He are compared with the available experimental data [183, 185] in Figure 7.9. As one can see, the LDA strongly overestimates the bonding, giving $\sim$ 15% smaller volume near $p = 15$ GPa than the experiment. This over-binding is reduced to below 5% at pressure above 100 GPa. At the same time, both gradient level approximations give results in good agreement with experiments. The correspondence between GGA and experiment is remarkable at pressure $\leq$ 30 GPa. For higher pressure LAG outperforms the other two approximations. However, because

[6] The low-pressure regime in rare gas solids is governed by van der Waals interactions, which cannot be described by the traditional density functional approaches. Therefore, the EMTO results for He are presented only for pressures above $\sim$ 10 GPa.

the experimental elastic constant of *hcp* He are available only for p =13–33 GPa [183], and for this pressure range the GGA gives the best agreement with the experimental pressure–volume data, the elastic constants have been computed using this approximation.

The complete set of hexagonal elastic constants of solid He is plotted as a function of pressure in Figure 7.10. We can see that the theoretical elastic constants agree very well with the experimental data [183]. In fact, the small discrepancies are below the typical errors obtained for the transition metals in conjunction with the GGA [17, 157, 188]. We find that all five elastic constant are positive and increase monotonously with p for the entire pressure interval considered in this study. Since the theoretical c_{ij} values from Figure 7.10 satisfy the stability conditions from Equation (6.25), the *hcp* phase of He remains mechanically stable up to at least 150 GPa.

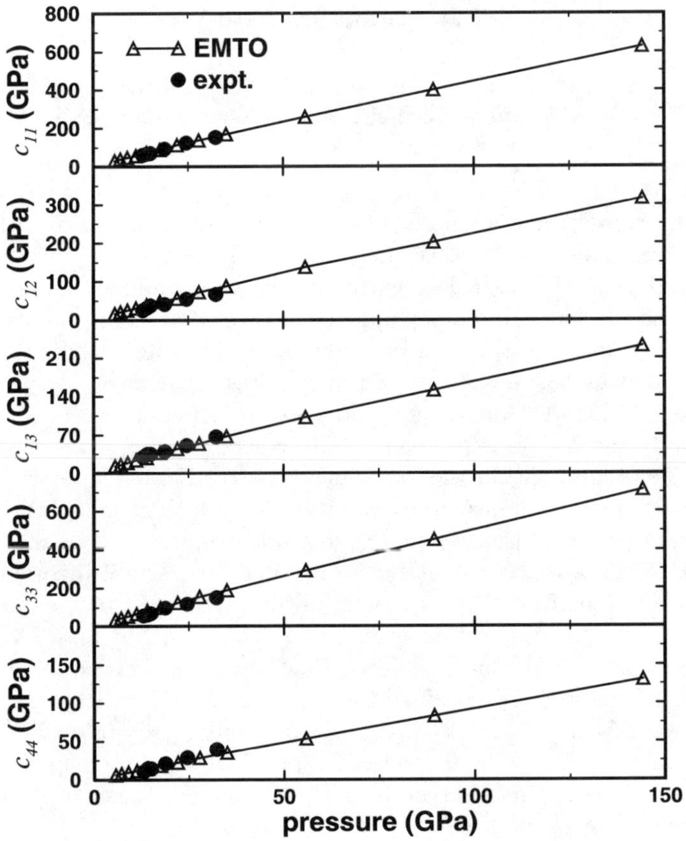

Fig. 7.10. Elastic constants of *hcp* He as functions of pressure. Open triangles: EMTO-GGA results, filled circles: experimental data [183].

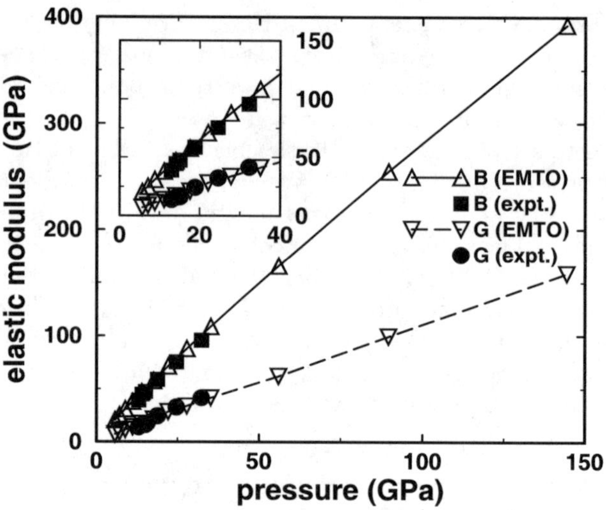

Fig. 7.11. Bulk and shear modulus of *hcp* He as a function of pressure. Open triangles: EMTO-GGA results, filled circles: experimental data [183]. In inset, data between 0 and 40 GPa are shown.

Within the accuracy of the EMTO method, the axial ratio of the hexagonal He is calculated to be ideal, i.e. $(c/a)_0 \approx 1.63$, and show negligible volume dependence. This is in line with experimental findings below 23.3 GPa [187]. Due to the flat volume dependence of $(c/a)_0$, at each pressure we have $c_{33}-c_{11} \approx c_{12}-c_{13}$, and thus the bulk modulus of *hcp* He reduces to Equation (6.58). The calculated anisotropy of the compressional wave, Δ_{P} (see Section 6.2) is around 1.12−1.13, and shows no pressure dependence. This result is in slight disagreement with experiment [183], where $c_{11} \approx c_{33}$ (*i.e.* $\Delta_{\mathrm{P}} \approx 1$) and $c_{12} \approx c_{13}$ was found within the experimental error bars. Therefore, the theoretical results violate the isotropy condition for a hexagonal symmetry [128]. The pressure factors of the two shear wave anisotropies, Δ_{S1} =1.60−1.66, and Δ_{S2} =0.87−0.83, are also calculated to be small. These anisotropy ratios are somewhat different from the experimental ones [183]. In particular, we find that the calculated Δ_{S1} for He is relatively close to that of solid H_2 [183]. It is interesting to note that the theoretical anisotropy ratios for *hcp* He are within the range of those obtained for hexagonal metals with $(c/a)_0 \approx 1.63$. For instance, Mg has $\Delta_{\mathrm{P}} = 1.04, \Delta_{\mathrm{S1}} = 1.19$, and $\Delta_{\mathrm{S2}} = 0.98$. These figures for Co are $1.17, 1.52, 1.06$ and for Re $1.12, 1.36, 0.95$ [189]. Therefore, at 0 K and zero pressure the monocrystalline He has similar anisotropy to the *hcp* metals from the periodic table.

In Figure 7.11, the theoretical bulk modulus and shear modulus of solid He are compared with the experimental values [183]. The polycrystalline shear modulus was calculated from the single-crystal elastic constants using the Voigt−Reuss−Hill averaging method (Section 6.3). The Debye temperature

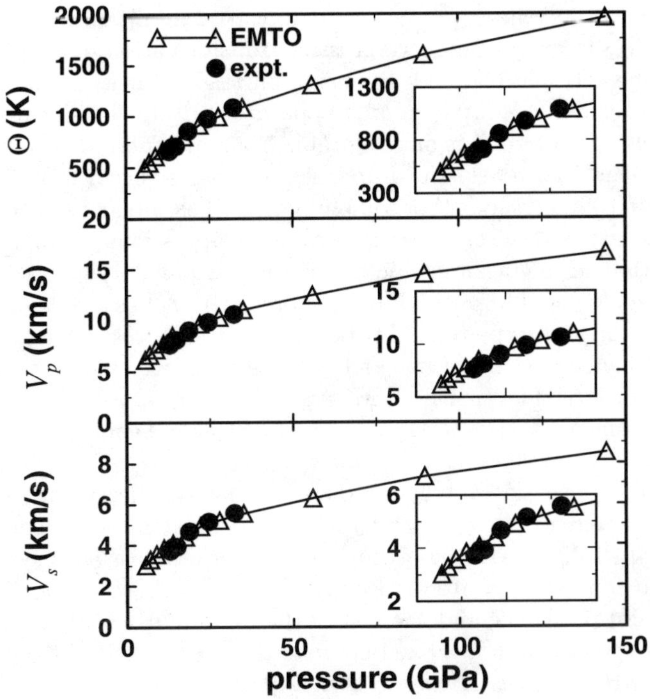

Fig. 7.12. Debye temperature (upper panel) and sound velocities (middle and lower panels) of *hcp* He as functions of pressure. Open symbols: EMTO-GGA results; filled circles: experimental data [183]. In insets, data between 0 and 40 GPa are shown.

θ_D, and the two sound velocities V_P and V_S, derived from the polycrystalline elastic moduli, are plotted on Figure 7.12. In both figures, we see a very good agreement between the theoretical and experimental data over the whole pressure range considered in the experiment.

The EMTO polycrystalline anisotropy ratio (Equation (6.91)) of *hcp* He is ~ 0.02, and shows negligible pressure dependence. Most of the cubic and low symmetry crystals have elastic anisotropy ratios between 0.0–0.21 [128, 134]. On this scale, the anisotropy of *hcp* He can be considered to be small. The low polycrystalline anisotropy ratio and its weak pressure dependence explain why solid He can be used as a quasi-hydrostatic medium in high-pressure experiments up to at least 150 GPa.

7.2.3 The Magnesium-silicate and Scandium-aluminate Perovskites

Magnesium-silicate perovskite is the most abundant mineral of the Earth's lower mantle, and thus the knowledge of its properties is crucial in advanced seismic research. Due to the marginal stability of $MgSiO_3$, a high temperature measurement of its elastic properties requires the simultaneous application of sufficient pressure to prevent retrogressive transformation to an amorphous or pyroxene phase [190]. The high frequency ultrasonic measurement cannot be reliably performed in a large volume apparatus within the silicate perovskite stability field, therefore, it is more suitable from an experimental point of view to measure the thermo-elastic properties of close structural analogues [191, 192, 193, 194]. Since the orthorhombic $ScAlO_3$ has molar volume, structural distortion relative to the cubic structure and compressional behavior [195, 196] similar to that of magnesium-silicate, it was proposed [194] to be the closest structural analogue of $MgSiO_3$. Furthermore, $ScAlO_3$ has also been considered frequently as a prototype of ABO_3 perovskite structures: its structure shows close resemblance to that of rare earth orthoferrites, *e.g.*, $GdFeO_3$ and $LuFeO_3$ [194]. Using the EMTO method, Magyari-Köpe *et al.* [197, 198] studied the thermodynamic properties of $MgSiO_3$ and $ScAlO_3$ and confirmed the observed similarities between these two perovskites.

The cubic perovskite structure has $Pm\bar{3}m$ symmetry and the Mg(Sc) cation is situated in the center of the cube defined by eight corner sharing $SiO_6(AlO_6)$ octahedra. In the orthorhombic phase with $Pbnm$ symmetry, the Mg(Sc) and four oxygen atoms occupy the ($4c$) positions, the Si(Al) atoms are in the ($4b$) positions, and the rest of the oxygen atoms are situated in the general ($8d$) positions. For a fixed volume, there are totally nine independent structural parameters in $Pbnm$ lattice. Experimental and theoretical investigations on $MgSiO_3$ and $ScAlO_3$ orthorhombic perovskites [190, 195, 197, 198] confirmed that the $Pbnm$ structural parameters remain nearly constant with increasing pressure. Because of that, the EMTO study was carried out using the experimental structural parameters [195, 199].

The EMTO ground state parameters of $ScAlO_3$ are listed in Table 7.6, along with the available experimental results for the orthorhombic phase. The mean deviations between the theoretical average atomic radius and bulk modulus and the experimental data [195, 193, 200, 201] are 0.4% and 6.5% for the LDA, and 2.0% and 7.9% for the GGA functionals, respectively. We also

Table 7.6. Theoretical ground state parameters of $ScAlO_3$ calculated using the EMTO method in combination with the LDA and GGA energy functionals. w, B and dB/dp denote the equilibrium average atomic radius (in Bohr), bulk modulus (in GPa) and pressure derivative of the bulk modulus, respectively. For comparison the available room temperature experimental data are also shown.

str.	w_{LDA}	w_{GGA}	$w_{\mathrm{expt.}}$	B_{LDA}	B_{GGA}	$B_{\mathrm{expt.}}$	$\left(\frac{dB}{dp}\right)_{\mathrm{LDA}}$	$\left(\frac{dB}{dp}\right)_{\mathrm{GGA}}$	$\left(\frac{dB}{dp}\right)_{\mathrm{expt.}}$
$Pbnm$	2.48	2.52	$2.47^{a,b}$	245	214	218^a,249^c 232^d	4.2	4.3	4.0^a
$Pm\bar{3}m$	2.49	2.53	-	228	198	-	4.2	4.3	-

^{a} [195], ^{b} [193], ^{c} [200], ^{d} [201].

Table 7.7. Theoretical (LDA) and experimental equilibrium atomic radii (w in Bohr), bulk moduli (B in GPa) and structural energy difference (ΔE in mRy/atom) of $MgSiO_3$ perovskite in orthorhombic ($Pbnm$) and cubic ($Pm\bar{3}m$) phases.

	EMTO			Full-potential			Experimental	
str.	w	B	ΔE	w	B	ΔE	w	B
$Pbnm$	2.358	253	-	2.349^a,2.333^b	266^a,259^b	-	2.357^b	261^b
$Pm\bar{3}m$	2.394	258	23.3	2.381^b	258^b	20^a,22.1^b	-	-

^{a} [202], ^{b} [203].

note that the theoretical pressure derivative of the bulk modulus for $MgSiO_3$, calculated within the LDA [202], is 4.2, which agrees very well with the LDA value from Table 7.6. The fact that for perovskites the LDA yields somewhat better equilibrium parameters than the GGA is consistent with our observation made in the case of semiconductors and insulators. Further support for this is given in inset of Figure 7.13, where the LDA pressure–volume relation for orthorhombic $ScAlO_3$ is compared with the room temperature experimen-

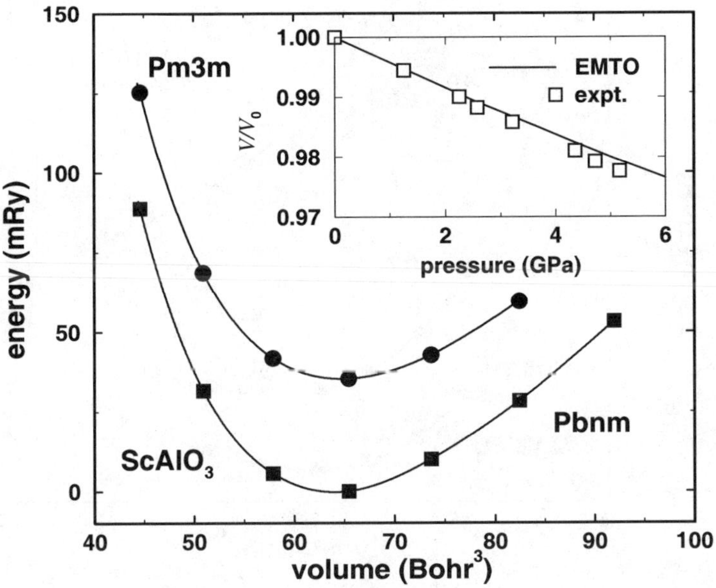

Fig. 7.13. Calculated total energies (per atom) of orthorhombic and cubic $ScAlO_3$ as a function of the average atomic volume. The energies are plotted relative to the orthorhombic ground state total energy. The symbols denote the EMTO total energies (LDA) and the connecting line is a Morse-type fit. In inset, the equation of state of orthorhombic phase is compared to the experimental results [195].

tal data [195]. There is an excellent agreement between the two sets of data, the average deviation between theoretical and experimental pressures being close to the numerical error bar of the EMTO calculation. In Table 7.7, the EMTO-LDA zero pressure volumes, bulk moduli and structural energy difference for $MgSiO_3$ are compared to the full-potential and experimental results [202, 203]. The agreement between the theoretical results, in view of the fact the full-potential techniques also have their own numerical approximations, may be considered satisfactory.

The EMTO-LDA total energies per atom for the *Pbnm* and $Pm\bar{3}m$ phases of $ScAlO_3$ and $MgSiO_3$ are plotted in Figures 7.13 and 7.14, respectively, as a function of the average atomic volume. The equations of state were obtained by fitting the LDA energy versus volume data to a Morse form (Section 6.1). It is seen from Figure 7.14, that for $MgSiO_3$ the stability of orthorhombic phase with respect to the cubic phase increases with pressure from 23.3 mRy/atom at zero pressure to $\sim$ 48 mRy/atom at 150 GPa. This is in line with the *ab initio* molecular dynamics results [203], where at 150 GPa an increase of $\sim$ 30 mRy/atom (relative to the zero pressure value) of the stability of the *Pbnm* phase with respect to the $Pm\bar{3}m$ phase was reported. On the other

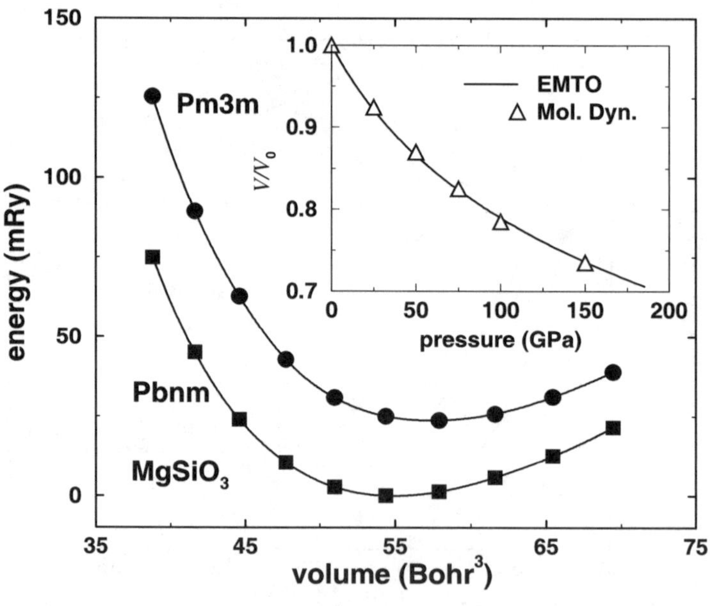

Fig. 7.14. Calculated total energies (per atom) of orthorhombic and cubic $MgSiO_3$ as a function of the average atomic volume. The energies are plotted relative to the orthorhombic ground state total energy. The symbols denote the EMTO total energies (LDA) and the connecting line is a Morse-type fit. In inset, the equation of state of orthorhombic phase is compared to the *ab initio* molecular dynamics results [203].

hand, for $ScAlO_3$ (Figure 7.13), the total energy difference between the two structures is ~ 34 mRy/atom and it shows a weak volume dependence. In the inset of Figure 7.14, we compare the EMTO pressure–volume relation for $MgSiO_3$ with the *ab initio* molecular dynamics results [203]. The two sets of data are very close, and the slightly increasing deviation between them at large pressures may be assigned to the lattice relaxation neglected in the EMTO study.

7.3 Transition-metal Nitrides

Transition-metal nitride films are frequently used as hard, protective coatings for soft materials. In Table 7.8, we give the EMTO results for the bulk equilibrium properties of titanium-nitride and vanadium-nitride. For comparison, a few GGA results calculated using pseudopotential [207, 208] and FP-LMTO [209] methods are also listed (numbers in parenthesis). Except the pseudopotential result for TiN, the gradient corrections seem to be insufficient to remove the strong LDA over-binding. Note that the GGA errors for w of TiN and VN are close to those obtained for elemental Ti and V (Table 7.1). Generally, when an approximation underestimates the volume it overestimates the bulk modulus and vice versa. Apparently, this is the case for VN, but both gradient-level bulk moduli for TiN are found to be smaller than the reported experimental value of 318 GPa [206]. Furthermore, in contrast to the experimental data, all three functionals give slightly larger bulk modulus for VN than for TiN. Since independent full-potential results [207, 208] are in line with the EMTO values, it is tempting to believe that the experimental B values for TiN and VN from

Table 7.8. Theoretical and experimental equilibrium atomic radii and bulk moduli of TiN and VN. The theoretical results were obtained using the EMTO method in combination with LDA [7], GGA [15] and LAG [17] energy functionals. The mean absolute values of the relative deviations between the EMTO values and the experimental data are given (Δ_{tot}). The equilibrium atomic radii (w) are expressed in Bohr and the bulk moduli (B) in GPa. Numbers in parenthesis are theoretical results calculated using pseudopotential [207, 208] and FP-LMTO [209] methods.

	str.	w_{LDA}	w_{GGA}	w_{LAG}	$w_{expt.}$	B_{LDA}	B_{GGA}	B_{LAG}	$B_{expt.}$
TiN	B1	2.433	2.480 (2.492[d],2.480[e])	2.459	2.486[a]	332	286 (278[d],270[e])	302	318[c]
VN	B1	2.346	2.398 (2.421[f],2.398[f])	2.373	2.427[b]	342	292 (317[f],330[f])	313	268[c]
Δ_{tot}		2.7%	0.7%	1.7%		16.0%	9.5%	11.0%	

[a] Expt. [204],
[b] Expt. [205],
[c] Expt. [206],
[d] Pseudopotential [207],
[e] FP-LMTO [209],
[f] Pseudopotential [208].

Table 7.8, measured for transition-metal nitride films on MgO substrate, do not reflect the correct bulk values.

For the above nitrides, the most stable surfaces are the (001) facets of the B1 structure [210]. This means that at equilibrium, the real TiN and VN surfaces are preferentially formed by the low-energy (001) facets. Using the EMTO method, we investigated the (001) surfaces of TiN and VN. In Table 7.9, the EMTO top layer relaxations are compared with the pseudopotential [207, 211, 212] and experimental [213] results. These EMTO calculations were carried out within the GGA. In the table, λ^b is the bulk inter-layer distance, and $\Delta\lambda^s_{\mathrm{Ti/V}}$ and $\Delta\lambda^s_{\mathrm{N}}$ denote displacements for the transition metals and nitrogen atoms from the surface layer with respect to the ideal geometry.

We find a rippled relaxation on both nitride surfaces. The Ti and V atoms move inward with 0.23 and 0.32 Bohr, respectively, which represent 5.5% and 8.5% of the theoretical equilibrium bulk inter-layer separations. On both nitride surfaces, the N atoms move outward with approximately 0.06 Bohr ($\sim$ 1%). The EMTO figures for the relaxation are in good accordance with other theoretical [207] and experimental [213] results. It is interesting that the crystallographic roughness[7] on the VN surface is 33% larger than on the TiN surface, *i.e.* the chemically flat VN surface is more corrugated in the crystallographic sense compared to the TiN surface.

In the absence of other all-electron results, the EMTO surface energies from Table 7.9 are compared with the available pseudopotential data [207, 211, 212]. We find that the EMTO surface energies are significantly larger than the corresponding pseudopotential values. It would be plausible to ascribe this discrepancy to the relaxation of the subsurface layers, which has been neglected in the EMTO study. However, the surface energies for unrelaxed

Table 7.9. Equilibrium surface geometry (distance relative to the ideal surface, $\Delta\lambda$ in Bohr) and the corresponding surface free energy (γ in J/m^2) for the (001) surfaces of TiN and VN. For comparison the bulk inter-layer distances (λ^b in Bohr) are also listed. The pseudopotential results [207, 211, 212] were obtained within the self-consistent GGA scheme. In the EMTO calculations, total energy was calculated within the GGA using the self-consistent LDA electron density.

surface		λ^b	$\Delta\lambda^s_{\mathrm{Ti/V}}$	$\Delta\lambda^s_{\mathrm{N}}$	$\gamma_{(001)}$
TiN	(001)	3.99	-0.23	0.06	1.98
			-0.23[1]	0.11[1]	1.36[1],1.28[3],1.36[4]
VN	(001)	3.86	-0.32	0.06	1.75
			-0.30[2]	0.02[2]	1.11[4]

[1] pseudopotential [207], [2] experiment [213],
[3] pseudopotential [211], [4] pseudopotential [212].

[7] By crystallographic roughness we mean the distance between the two monoatomic top layers: a transition metal and a N layer.

geometries also show large discrepancies. For instance, in EMTO the top layer relaxation reduces the surface energy of TiN by 0.32 J/m^2, whereas this figure in pseudopotential study varies between 0.28 J/m^2 [207] and 0.47 J/m^2 [211]. Hence, the observed deviations in the surface energies for nitrides should have another origin.

Liu *et al.* [212] calculated the surface energies of several ceramics with B1 structure using a pseudopotential approach combined with both LDA and GGA functionals. For all ceramics considered, the surface energies turned out to be very sensitive to the choice of exchange-correlation functional. For instance, their GGA surface energy for TiN agrees well with the previous pseudopotential results from Table 7.9, whereas their LDA surface energy is 2.18 J/m^2, *i.e.* $\sim$ 60% larger than the GGA value. The latter is in fact close to the EMTO value of 1.98 J/m^2. We recall that the EMTO self-consistent calculations for TiN and VN were performed within the LDA for the effective potential and only the total energy was computed within the GGA. This may explain why the EMTO surface energies from Table 7.9 are significantly higher compared to the GGA pseudopotential values.

Before closing this section, we comment on the accuracy of the LDA and GGA for metal surfaces. The performance of some common exchange-correlation approximations was studied within the self-consistent jellium surface model by Perdew *et al.* [9]. They found that the LDA gives surface energy in good agrement with the experimental result, especially for slowly varying density profiles. Its success was ascribed to a cancelation between the errors in the exchange and correlation energies. The gradient correction to the LDA [15] represents an important improvement for the correlation part, but it underestimates the exchange energy, and as a consequence it gives surface energies which are 7$-$16% lower than the LDA values for jellium and 16$-$29% lower than the experimental results. Another important observation is that the LDA yields surface energies in better agreement with the broken bond model than the GGA [212]. The broken bond model is based on the cohesive energy[8], which can be calculated accurately within the GGA. Assuming that the broken bond model is valid for nitride surfaces, one concludes that the GGA surface energies are not reliable. On these grounds, one may therefore prefer the LDA approximation for the effective potential near the metal surfaces.

[8] According to the broken bond model, the surface energy per atom is $\left(2-\eta-\sqrt{\eta})\right) E_c/2$, where η is the ration between the coordination numbers of surface and bulk atoms, and E_c is the cohesive energy [212].

8

Binary Alloys

In this chapter the application of the EMTO method in the case of binary substitutional random alloys will be demonstrated. First, in Section 8.1, we shall describe how the main shortcoming of the single-site approximation is handled within the present method. In Section 8.2, applications in the case of Al-based alloys and $(MgAl)B_2$ will be presented. The thermodynamic and mechanic stability of two well-known Hume–Rothery systems will be investigated in Section 8.3. Theoretical studies of some binary transition metal alloys will be reviewed in Section 8.4. The application of the EMTO method to f-electron systems will be illustrated on CeTh alloy in Section 8.5. Finally, in Section 8.6, the equilibrium concentration profile calculation will be illustrated in the case of the PdAg system. A few important numerical details of the EMTO calculations presented in this chapter are listed in Section C.2.

8.1 The Single-site Approximation

Within the *Coherent Potential Approximation* (CPA), one single impurity is embedded into a homogeneous effective medium described by the coherent potential and coherent Green function. Accordingly, no information is provided on how the actual potential is altered around the impurity. This single-site approximation is the main drawback of the CPA. In the Schrödinger equation, the single-site approximation is overcome by compensating the charge redistribution from the effective medium. However, in the Poisson equation no such simple solution exists. Today, probably the most efficient approach to handle this problem is the screened impurity model (SIM) introduced by Korzhavyi *et al.* [121, 122]. This model involves two *a priori* parameters, which are determined from supercell calculations. They are obtained by matching the CPA total energy and charge transfer to the real total energy and charge transfer calculated using a large unit cell. In many applications, these two parameters are related to each other and thus can be chosen to be the same.

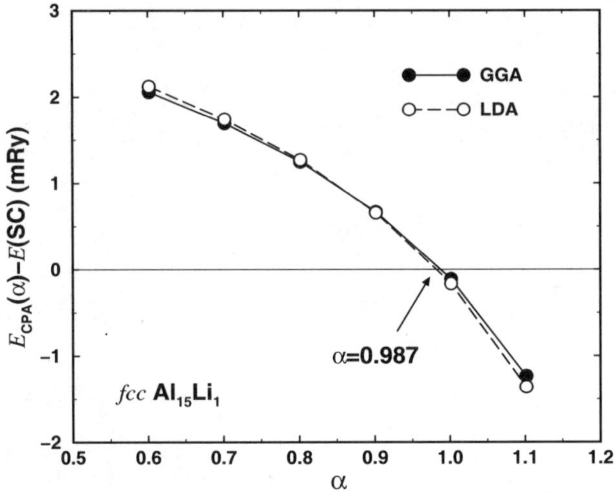

Fig. 8.1. Total energy for *fcc* $Al_{0.9375}Li_{0.0625}$ random alloy calculated using the EMTO method in combination with the LDA (open circles) and the GGA (filled circles) as a function of SIM parameter α [121, 122]. The energies are plotted relative to the total energy of an *fcc* based supercell containing 15 Al and 1 Li.

The EMTO results from Chapters 8 and 9 were calculated using a single SIM parameter α.

We illustrate the way the α is determined in the case of Al-rich AlLi, Pu-rich PuGa and PuAl alloys. At ambient conditions, the dilute AlLi solid solutions have the *fcc* crystal structure. For Pu-based alloys, we consider the δ-Pu phase (*fcc*), which is stable at temperatures between 593 and 736 K. First, we set up a 16-atom supercell (SC) containing 15 Al(Pu) atoms and one impurity, corresponding to an average impurity concentration of 6.25%. The total energy of the *fcc* supercell (E(SC)) was computed using the EMTO method. Special care was taken on the accuracy of the SC calculations (*e.g.*, using a well converged Brillouin zone sampling). In the second step, using the CPA, we computed the total energies of $Al_{1-x}Li_x$, $Pu_{1-x}Ga_x$ and $Pu_{1-x}Al_x$ systems for $x = 0.0625$ as a function of α. In both SC and CPA calculations the lattice constant was fixed to the bulk lattice constant calculated for elemental Al(Pu). The CPA energies are plotted in Figures 8.1 and 8.2 relative to the total energy per atom calculated for the corresponding SC. Results are shown for LDA and GGA energy functionals.

In the AlLi system, the considerable CPA charge misfit leads to a CPA total energy $E_{\mathrm{CPA}}(\alpha)$ which sensitively depends on α. The ~ 2 mRy difference between the CPA and SC total energies, calculated for $\alpha = 0.6$ rapidly drops to zero and becomes negative with increasing α. The optimal SIM parameter is obtained from the condition $E_{\mathrm{CPA}}(\alpha) - E(\mathrm{SC}) = 0$. For the GGA energy

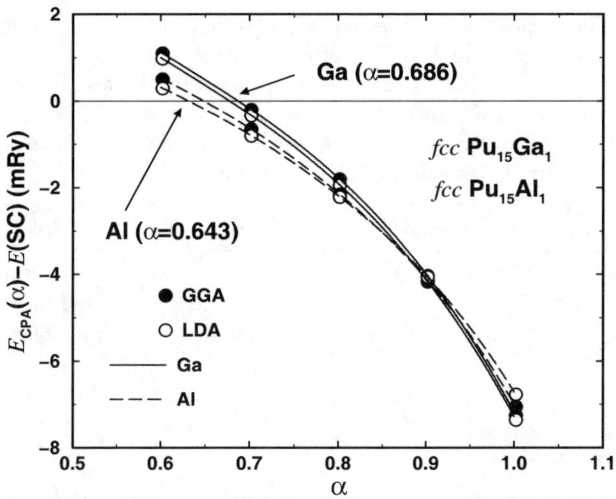

Fig. 8.2. Total energy for *fcc* $Pu_{0.9375}Ga_{0.0625}$ (solid lines) and $Pu_{0.9375}Al_{0.0625}$ (dashed lines) random alloys calculated using the EMTO method in combination with the LDA (open circles) and the GGA (filled circles) as a function of SIM parameter α [121, 122]. The energies are plotted relative to the total energies of *fcc* based supercells containing 15 Pu and 1 Ga or Al, respectively.

functional, this is realized for $\alpha_{\mathrm{AlLi}} = 0.987$. Note that the optimal α is not sensitive to the choice of exchange-correlation functional.

In Pu-based alloys, the optimal SIM parameters turn out to be significantly lower compared to α_{AlLi}. From Figure 8.2, we have $\alpha_{\mathrm{PuGa}} = 0.686$ and $\alpha_{\mathrm{PuAl}} = 0.643$, and, similar to AlLi, they exhibit a weak dependence on the energy functional. Most of the binary alloys have optimal α close to those obtained for the above Pu-based alloys, and only a few systems require SIM parameters as large as 1.

There are situations, like in Al-rich AlMg alloys, where the CPA charge misfit (defined in Equation (5.22)) is small and therefore the CPA total energy shows negligible dependence on the SIM parameter. In this situation, the above technique is inappropriate for finding a proper α value. However, for these systems the calculated physical parameters are also less sensitive to the choice of α.

8.2 Light Metal Alloys

8.2.1 Aluminum–Lithium Alloys

For years, Al-based alloys were in the focus of theoretical and experimental investigation. Their complex electronic structure and the observed anomalous

Table 8.1. Theoretical and experimental elastic constants (in GPa) and elastic anisotropy for *fcc* Al. The EMTO values were obtained within LDA, GGA and LAG approximations for the exchange-correlation functionals. References are given for the full-potential (FP) and experimental (expt.) data. The linear augmented plane wave (LAPW) study [224] was carried out for the experimental volume, and the experimental values from Ref. [223] were extrapolated to 0 K.

	LDA	GGA	LAG	FP	expt.
c_{11}	110.8	98.9	101.3	121.9±1.6[a], 101.5[b] 110.5[c], 103.3[d]	108[e] 106.9[f] 114.3[g]
c_{12}	72.8	65.7	66.9	62.7±1.3[a], 70.4[b] 58.0[c], 53.3[d]	61[e], 60.8[f] 61.9[g]
c_{44}	45.1	38.1	39.6	38.4±3.0[a], 31.7[b] 31.1[c], 28.5[d]	29[e], 28.2[f] 31.6[g]
$A_{\rm E}$	-0.79	-0.71	-0.73	-0.21[a], -0.46[b] -0.12[c], -0.09[d]	-0.13[e,f,g]

[a] LAPW, LDA, Ref. [221]. [e] Ref. [219].
[b] LMTO, LDA, Ref. [222]. [f] Ref. [225].
[c] LMTO, GGA, Ref. [223]. [g] Ref. [223].
[d] LAPW, LDA, Ref. [224].

trends of bulk properties against composition have always presented a great scientific challenge. For instance, several theoretical investigations [121, 214, 215, 216, 217], based on the virtual crystal approximation (VCA), CPA and cluster expansion approach, were dedicated to reveal the origin of the rather unusual properties of AlLi. As two interesting features of this system, one should mention the contraction of the equilibrium volume relative to a linear interpolation between Al and Li [218], and the drastic increase of the Young's modulus with Li content [219]. Although most of the above studies reproduced the characteristic features of the composition dependent equilibrium volume, the theoretical mapping of the elastic properties against concentration has remained a problem. Taga *et al.* [220], using the EMTO method, have reviewed this question and presented a detailed theoretical description of the ground state properties of AlLi solid solutions.

In order to evaluate the relative merits of the LDA, GGA and LAG approximations in the case of Al-rich Al-Li alloys, we compare their performances for the equilibrium volume and elastic properties of *fcc* Al. In Section 7.1.1 we saw that for the equilibrium volume and bulk modulus of Al the GGA is superior to that of the LDA and LAG approximations. In Table 8.1, we compare the EMTO single-crystal elastic constants for *fcc* Al with the experimental data [219, 225] and some former *ab initio* results calculated using full-potential (FP) linear augmented plane wave (LAPW) [221, 224] and linear muffin-tin orbitals (LMTO) [222, 223] methods. Again, letting the error connected with these calculations be described by the difference between the FP

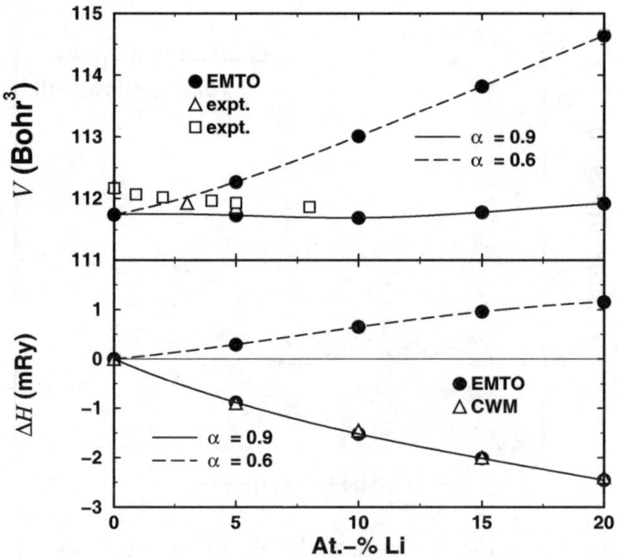

Fig. 8.3. Concentration dependence of the theoretical and experimental equilibrium atomic volume (upper panel) and mixing enthalpy (lower panel) of AlLi random alloys. Experimental atomic volumes are from Ref. [218] (triangles) and Ref. [226] (squares). CWM stands for the results obtained using the Connolly–Williams method [121]. The two sets of EMTO results correspond to two different SIM parameters from Equations (5.22) and (5.33): $\alpha = 0.9$ (solid line) and $\alpha = 0.6$ (dashed line).

results from Table 8.1, we obtain that the agreement between the EMTO and former theoretical values is very good. The calculated average deviations between the experimental [219] and EMTO data for elastic constants, obtained within LDA, GGA and LAG, are 25%, 10%, and 13%, respectively. Therefore, we can conclude that in general the GGA yields significantly better ground state properties for Al compared to the LDA, and marginally better compared to the LAG approximation. The rest of the results presented in this section have been obtained within the GGA.

In Figure 8.3, we give the EMTO equilibrium volume and enthalpy of formation for AlLi alloys as a function of concentration. Data are shown for two different α values from Equations (5.22) and (5.33). We find that $\alpha \approx 0.9$, which is close to the optimal value determined in the previous section, reproduces well the observed trend in the equilibrium volume, whereas $\alpha = 0.6$ gives an increase in V with Li addition, which is in between the experimental value and the one estimated from the linear rule-of-mixture. The enthalpy of formation of fcc $Al_{1-x}Li_x$ alloy is calculated as

$$\Delta H(x) = E(\mathrm{Al}_{1-x}\mathrm{Li}_x) - (1-x)E(\mathrm{Al}) - xE(\mathrm{Li}), \quad (8.1)$$

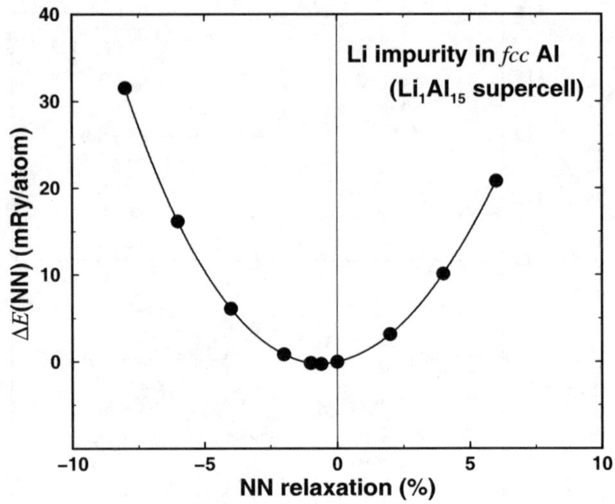

Fig. 8.4. Variation of the total energy of $Al_{15}Li_1$ supercell as a function of the nearest-neighbor (NN) Al–Li distance. The energy minimum corresponds to -0.6% inward relaxation of the first coordination shell around the Li impurity.

where all the energies are obtained for the theoretical equilibrium volumes and expressed per atom. $E(\text{Al})$ and $E(\text{Li})$ are the total energies of fcc Al and Li, respectively. In Figure 8.3 the EMTO enthalpy of formation is compared with that obtained using the Connolly–Williams method (CWM) [121]. Within the CWM [66] the Madelung energy is treated exactly, and thus gives a good reference to establish the accuracy of our approach for the formation energy of completely random alloys. The perfect agreement between values calculated using CWM and the EMTO method with $\alpha = 0.9$ demonstrates that the charge-transfer effects can be adequately taken into account in the single-site EMTO approach. The EMTO result for $\Delta H(x)$ confirms the observations [121, 214] that the thermodynamic stability of AlLi solid solutions is to a large extent determined by the Madelung energy accounting for the charge-transfer between Al and Li subsystems.

Using a supercell approach, one can study how the local lattice relaxation (LLR) influences the heat of formation of AlLi solid solutions. The EMTO equilibrium volume of a 16-atom supercell containing 15 Al and 1 Li is 111.59 Bohr^3. This value is very close to 111.73 Bohr^3 calculated for $Al_{0.9375}Li_{0.0625}$ random alloy using the CPA with $\alpha = 0.9$. Figure 8.4 shows the variation of the total energy of $Al_{15}Li_1$ (ΔE) as a function of the nearest-neighbor (NN) Al–Li distance. From the energy minimum we find approximately -0.6% NN relaxation, *i.e.*, in $Al_{15}Li_1$ the Al–Li distance decreases by 0.6% compared to the equilibrium Al–Al bond length in pure Al. The LLR decreases the total energy relative to the unrelaxed structure by -0.242 mRy/atom. Compared to other alloys, the LLR in AlLi can be considered very small. Ruban *et al.*

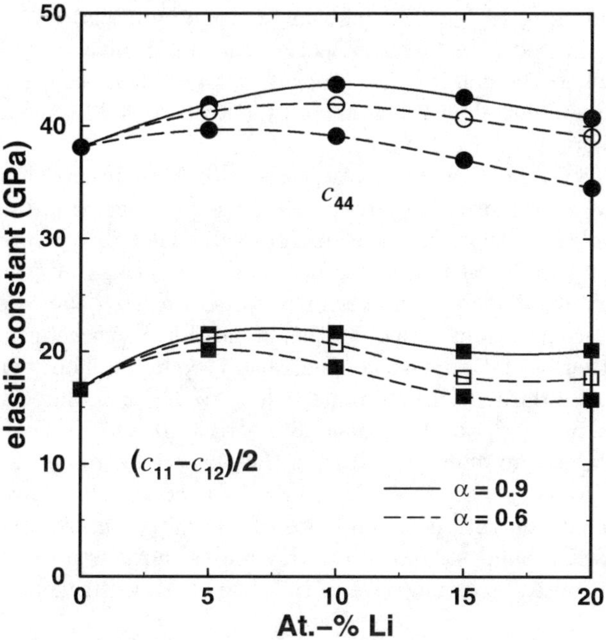

Fig. 8.5. Theoretical (EMTO-GGA) cubic shear moduli $(c_{11}-c_{12})/2$ and c_{44} of AlLi alloys as a function of Li content. Filled symbols correspond to the results obtained using $\alpha = 0.9$ (solid line) and $\alpha = 0.6$ (dashed line) in Equations (5.22) and (5.33). Open symbols denote results calculated at volumes corresponding to $\alpha = 0.9$, and using $\alpha = 0.6$ for the elastic constant calculations.

[227] proposed that the LLR is mainly governed by the change of the volume of the host material. This is in line with the small difference obtained ($\sim 0.2\%$) between the theoretical equilibrium volumes of $Al_{15}Li_1$ and *fcc* Al.

The enthalpy of formation for a 16-atom supercell is -1.588 mRy/atom for an ideal *fcc* underlying lattice and -1.830 meV/atom for relaxed structure. These figures slightly exceed the mixing enthalpy obtained for the $Al_{0.9375}Li_{0.0625}$ random alloy using $\alpha = 0.9$ (Figure 8.3). Note that the optimal $\alpha \approx 0.987$ would reproduce exactly the supercell result.

In Figure 8.5, we illustrate the CPA charge-transfer effects on the single-crystal elastic moduli of AlLi alloys. The two sets of results for c' and c_{44}, marked by filled symbols connected with continuous and dashed lines, were obtained from self-consistent EMTO calculations using $\alpha = 0.9$ and $\alpha = 0.6$, respectively. The variation of both sets of elastic constants with Li content is smooth. They exhibit similar concentration dependencies, and the only important difference between them is the position of the maxima. We find that the maximum values in c' and c_{44} are shifted towards higher concentrations with increasing α. When the calculations are carried out at fixed volumes, *e.g.*

those corresponding to $\alpha = 0.9$ (shown by open symbols), the effect of α is even less pronounced. The largest effect on the cubic shear moduli is obtained for $x = 0.2$, where we get $\partial \log c_{44}/\partial \log \alpha \approx 0.45$. This variation is one order of magnitude smaller than $\partial \log \Delta H(0.2)/\partial \log \alpha \approx 4.42$, calculated for the enthalpy of formation (Figure 8.4).

In Figure 8.6, the EMTO elastic constants for AlLi random alloys, obtained using $\alpha = 0.9$, are compared with the experimental data by Müller *et al.* [219]. We observe that the experimental value is slightly overestimated for c_{12} and c_{44} and underestimated for c_{11}. Such deviations are, however, typically obtained for elemental metals in conjunction with the existing density functional approximations. On the other hand, for all three elastic constants we find that their variation with concentration are in excellent agreement with the experimental data. At Li contents below 5% the calculated changes with concentration in c_{11}, c_{12} and c_{44} are 0.86, −1.13 and 0.77 GPa per At.-% Li, respectively. These numbers are close to the observed average variations 0.33, −0.95, and 0.51 GPa per At.-% Li [219]. Taga *et al.* [220] have shown that the nonlinear concentration dependence of the cubic elastic constants arises from the peculiar band structure of AlLi alloys. For instance, the band energy contribution to c_{44} is negative in the case of Al, positive for intermediate

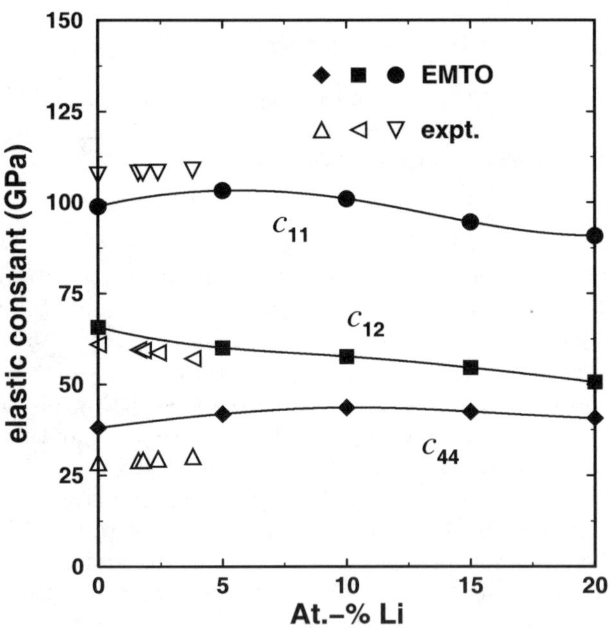

Fig. 8.6. Comparison between theoretical (EMTO) and experimental [219] single-crystal elastic constants for AlLi random alloys.

(approx. 5–15%) Li concentrations, and zero for $\sim 20\%$ Li, which results in a maximum in c_{44} near 10 At.-% Li.

We estimate the impact of local lattice relaxations on the single-crystal elastic constant by computing the tetragonal shear modulus c' of $Al_{15}Li_1$ supercell with and without NN relaxation. The agreement between the non-relaxed supercell value of 20.3 GPa and 21.2 GPa obtained for $Al_{0.95}Li_{0.05}$ random alloy (Figure 8.6), is satisfactory, especially if one takes into account the numerical difficulties associated with elastic constant calculations. The relaxed supercell tetragonal elastic constant is 20.4 GPa. Within the error bars of the EMTO calculations, this value is identical to that obtained for the unrelaxed geometry. The almost vanishing effect of the LLR on the elastic constant of AlLi solid solutions can be ascribed to the small volume change on alloying. However, in systems where the lattice relaxation is more pronounced, like in CuAu alloys, a substantially larger impact of LLR on the elastic properties can be expected.

In the variation of the elastic constants c_{ij} with the concentration x of solute atoms in Figure 8.6, we may single out that part, which can be accounted for as due to an average change in the volume, *i.e.* in the lattice parameter. Data for higher-order elastic constants of Al give the pressure dependence $\partial c_{11}/\partial p = 5.9$, $\partial c_{12}/\partial p = 3.3$, $\partial c_{44}/\partial p = 1.9$ [128]. From experiments [226] on the lattice parameter a of dilute AlLi alloys we get $(1/a)(\partial a/\partial x) = -0.011$. When combined with the bulk modulus of Al, we get $\partial c_{11}/\partial x = 14.4$ GPa, $\partial c_{12}/\partial x = 8.1$ GPa, $\partial c_{44}/\partial x = 4.6$ GPa. Thus, the effect of alloying on the lattice parameter would account for about half of the increase observed in c_{11} (~ 33 GPa [219]), about 1/10 of the increase in c_{44} (~ 51 GPa [219]), but it has a sign opposite to that observed for c_{12} (*cf.* Figure 8.6). It follows that the Li solute atoms have an influence on c_{12} that depends crucially on the changes in the electronic structure.

The theoretical polycrystalline elastic moduli in Figure 8.7 (solid lines) were calculated using single-crystal results from Figure 8.6 and the averaging techniques presented in Section 6.3.1. In figure, we included the experimental data on AlLi alloys by Müller [219], and those on commercial 2024 Aluminum Alloy by Sankaran and Grant [228]. The observed decrease of the bulk modulus, and the sharp increase of the Young's and shear moduli at low Li concentrations, are well reproduced by the theory. In order to illustrate how sensitive the polycrystalline elastic moduli are to the value of α, in Figure 8.7 the theoretical values obtained for $\alpha = 0.6$ are also shown (dashed lines). The small effect of α on the cubic elastic constants demonstrated in Figure 8.5, can be evidenced also in the case of B and G. A somewhat larger effect is obtained for the Young's modulus, where the experimental value for 11.4% Li is poorly reproduced by the theoretical curve obtained for $\alpha = 0.6$. However, it is not clear whether the 18.3% Young's modulus enhancement in this commercial alloy, relative to that of pure Al, is due to the solid solution itself or to the intermetallic phase, which forms within the solid solution matrix above $\sim 12\%$ Li [228].

The ratio between the bulk modulus and the shear modulus, shown in Figure 8.7 as B/G, is a measure of the ductility of solids: ductile alloys are characterized by large B/G ratios, whereas low B/G ratios are representative of brittle solids [229]. We find that a small amount of Li makes the alloy more brittle compared to pure Al. In dilute AlLi alloys the calculated B/G decreases with 9.6% per At.-% Li, compared to the experimental decrease of 7.6%. We note that the opposite trend for c_{12} from Figure 8.6, compared to c_{11}, leads to a rapid increase in the cubic shear modulus $c' = (c_{11} - c_{12})/2$ and thus is essential for the observed rapid decrease in B/G on alloying Al with Li.

8.2.2 Other Aluminum-based Alloys

Following the procedure presented in the previous section, one can compute the ground state bulk properties of any substitutional random binary system.

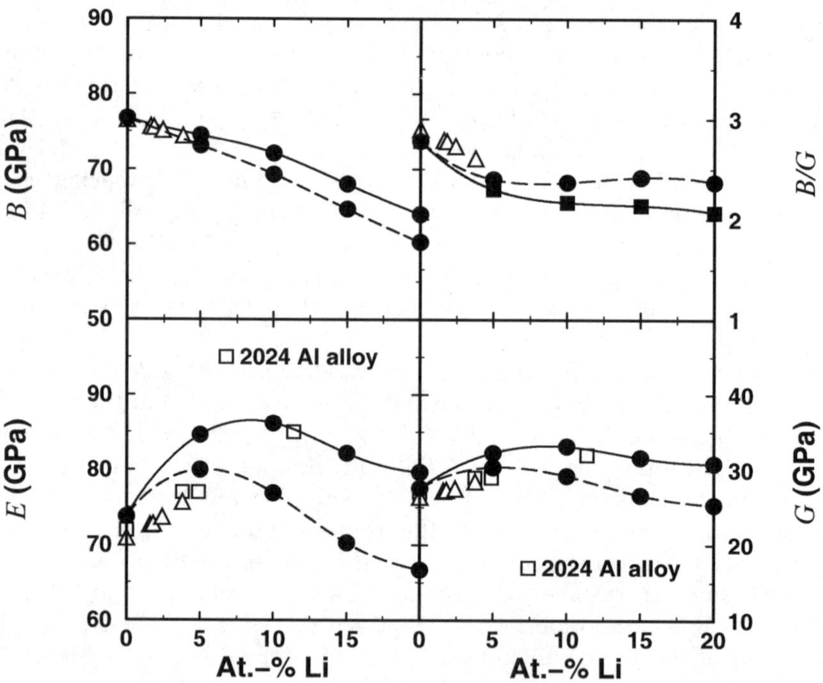

Fig. 8.7. Comparison between the EMTO (filled circles) and experimental (open triangles) [219] polycrystalline elastic moduli (B, E, and G stand for bulk, Young's, and shear modulus, respectively) of AlLi random alloys. Continuous and dashed lines correspond to the two sets of self-consistent EMTO results from Figure 8.5. For reference we also included experimental data for the Young's and shear moduli (open squares) obtained for the 2024 commercial aluminum alloy [228].

Below we quote some EMTO results obtained for Al-rich AlCu, AlMg, AlMn, AlSi and AlZn solid solutions. In Figure 8.8, the theoretical single-crystal elastic constants for AlMg are compared to the available experimental data [230]. We find that the variations of the theoretical values with concentration are in perfect agreement with the experimental data. In particular, we point out that both the experimental and theoretical c_{11} and c_{12} decrease whereas c_{44} slightly increases with Mg addition. Note that, in contrast to the effect of Li addition (Figure 8.6), alloying with Mg decreases c_{11} of Al. It is gratifying that the EMTO method correctly accounts for such interesting differences between alloying elements.

The calculated size misfit parameters ε_b (Equation (6.111)) for five Al-based solid solutions are compared to the experimental values in Figure 8.9. An excellent agreement between the computed and experimental ε_b values is observed. This figure also shows the calculated elastic misfit parameter ε_G as a function of theoretical ε_b. We observe certain correlation between ε_b and ε_G. For example, both the volume and elastic constant changes are negligible upon Zn addition. Elements producing large $|\varepsilon_b|$ also lead to nonzero ε_G. According to the theoretical data from Figure 8.9, we can see that the elastic misfit for Mn, Cu and Mg contributes by $\sim$ 25% to $\varepsilon_L^{4/3}$, *i.e.* to solid solution hardening (Section 6.6), whereas this effect is below 3% in the case of Si and Zn.

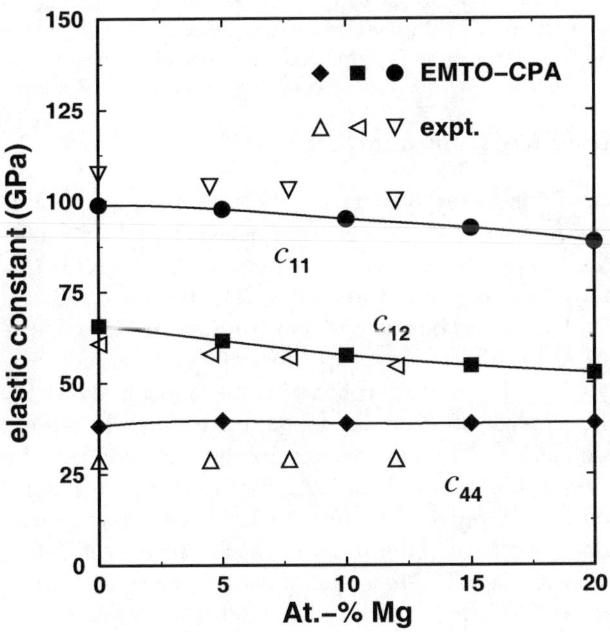

Fig. 8.8. Comparison between theoretical (EMTO) and experimental [230] single-crystal elastic constants for AlMg random alloys.

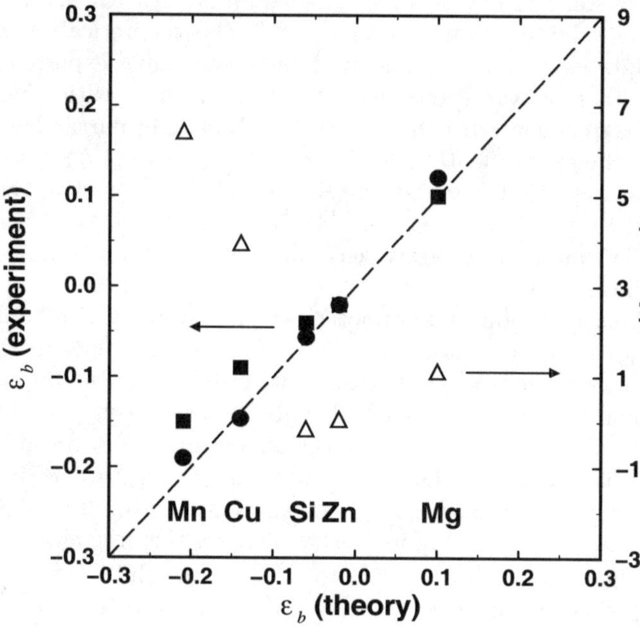

Fig. 8.9. Misfit parameters for selected Al-based alloys (alloying elements shown at the bottom of the figure). Left axis: experimental ε_b versus theoretical ε_b (Ref. [231]: circles, Ref. [232]: squares); right axis: theoretical ε_G versus theoretical ε_b.

8.2.3 Magnesium–Aluminum Diboride

The discovery of superconductivity in MgB_2 ceramic compound placed this material in the focus of several research activities [88, 89, 90]. The crystal structure of MgB_2 has the hexagonal symmetry (space group $P6/mmm$) with $c/a = 1.14$. Layers of Mg and B atoms are located at $z = 0$ and $z = 0.5c/a$, respectively. It has been shown [89] that additional electrons to MgB_2, *e.g.* substitution of Al for Mg, results in the loss of superconductivity. This transition is associated with structural collapse of the hexagonal lattice in c direction [89]. The rapid decrease of c/a in Al doped magnesium diboride is, in turn, related to the smaller ($\sim 7\%$) c-lattice spacing in AlB_2 compared to that of the MgB_2. The crystal structure of $Mg_{1-x}Al_xB_2$ random alloy can be investigated using the EMTO method. In Figure 8.10, we show the calculated structural parameters along with the experimental results from Ref. [89] as a function of Al content. The overall agreement between theoretical and experimental data is very good. Both methods predict almost constant a and decreasing c lattice parameters with increasing Al content. The EMTO method partially captures the small discontinuity in the observed c-lattice spacing: there is a clear change in the slope of theoretical c at about 20% Al. However, for a more

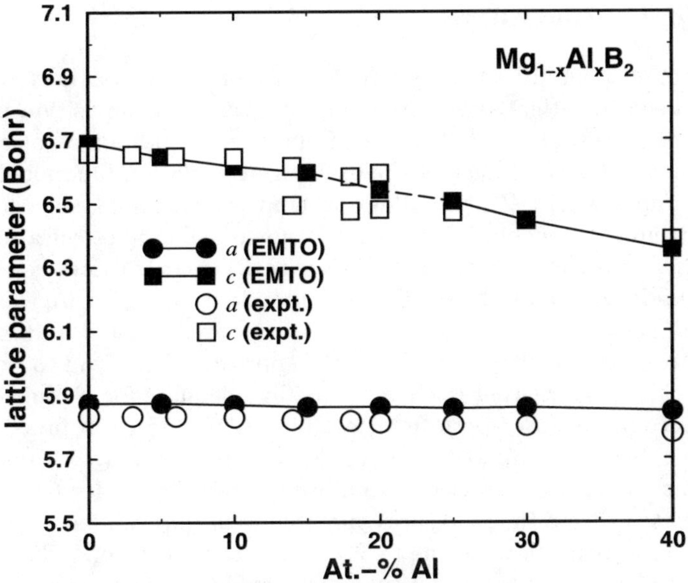

Fig. 8.10. The composition dependence of the EMTO (closed symbols) and experimental [89] (open symbols) lattice parameters in Al doped magnesium diboride, $Mg_{1-x}Al_xB_2$.

complete study of the structural collapse in $Mg_{1-x}Al_xB_2$ system, the possible formation of pure Al and Mg layers should also be taken into account.

8.3 Hume–Rothery Alloys

The Hume–Rothery binary alloys present classical systems for testing new alloy theories [59, 78, 108, 115, 233]. Here we consider two common systems: the CuZn and AgZn alloys. Due to their ability of cold working, outstanding aqueous corrosion resistance, and also to the pleasant lustrous appearance, the CuZn alloys or brasses have been used for hundreds of years. The AgZn alloys find their application in rechargeable batteries with very high energy density. According to the phase diagrams of $Cu_{1-x}Zn_x$ [232] and $Ag_{1-x}Zn_x$ [234], in both systems the *fcc* α-phase is stable for $x \leq 0.37$, and near equimolar concentrations the *bcc* β-phase is stabilized. In the CuZn system the *hcp* ϵ-phase is stable for $0.78 \leq x \leq 0.86$, while in the AgZn system this interval is somewhat larger, $0.68 \leq x \leq 0.87$. Above $\sim$ 97% Zn in both systems the second hexagonal structure, the η-phase, is formed.

8.3.1 Copper–Zinc Alloys

The electronic structure and elastic properties of Hume–Rothery binary alloys have been investigated as a function of chemical composition using the EMTO method [78, 115, 116, 235]. One fundamental question associated with such calculations is the choice of the exchange-correlation functional. This is especially important for Zn-rich alloys, where the hexagonal axial ratio shows a strong volume dependence. Usually, it is not possible to select a functional that has the same accuracy for the whole range of concentrations. For instance, the LDA underestimates the equilibrium atomic radii of Cu, Ag and Zn by 2.6%, 1.6% and 2.6%, whereas the GGA overestimates them by 0.8%, 1.7% and 0.6%, respectively. It is seen that both approximations lead to systematic errors for CuZn, but neither of them are really adequate for AgZn. Note that the difference between the performances of the LDA and GGA functionals for the end members is even more pronounced in AuZn alloys (see Table 7.1).

In Figure 8.11, the theoretical equilibrium atomic radii for $Cu_{1-x}Zn_x$ alloys, calculated within the LDA, are compared with the available experimental data. The agreement between the two sets of data is very good. The errors in the α and η phases are practically constant. Somewhat larger errors are found for the ϵ phase, but even here the concentration dependence is negligible. The equilibrium atomic radii for the two *hcp* lattices were determined at the theoretical equilibrium hexagonal axial ratios. These are plotted in Figure 8.12, where for comparison the experimental c/a values [232] are also shown. The anomalously large axial ratio of pure Zn is slightly underestimated, which is

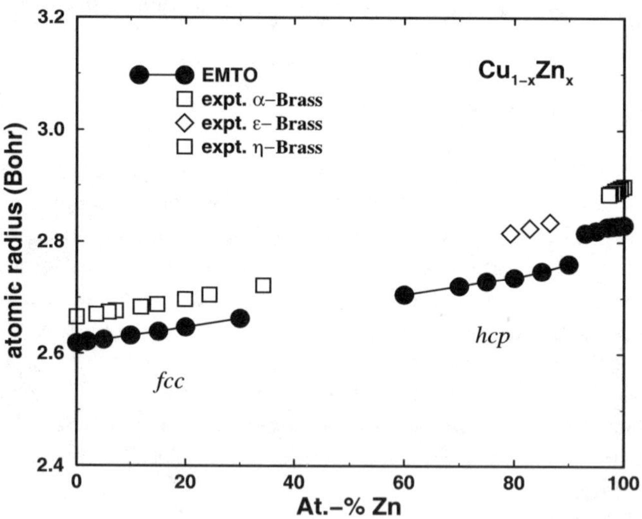

Fig. 8.11. Theoretical (EMTO-LDA) and experimental [189] equilibrium atomic radius of $Cu_{1-x}Zn_x$ random alloy as a function of Zn content.

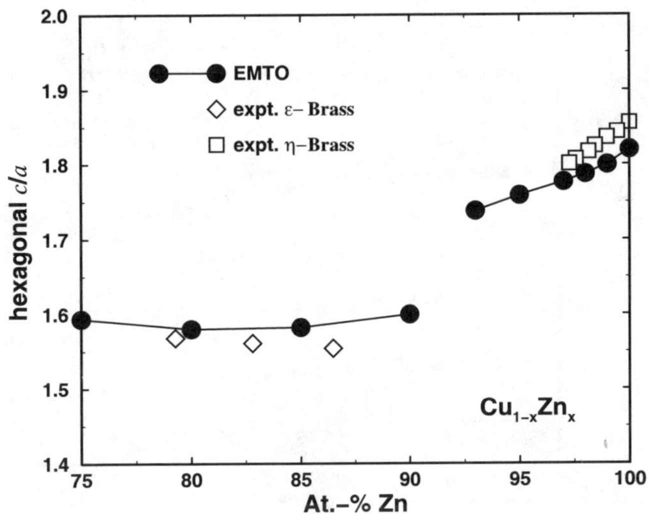

Fig. 8.12. Comparison between the theoretical (EMTO-LDA) and experimental [232] hexagonal axial ratio for ϵ- and η-brass. The axial ratio is plotted as a function of Zn concentration.

due to the LDA over-binding. On the other hand, the theoretical c/a for Zn is calculated to be reduced by 2.4% at 3 At.-% Cu, in very good agreement with the experimental result. At low Cu concentrations the η-phase having a large c/a ratio is found to be the ground state structure of the CuZn alloy. However, with increasing Cu concentration a second total energy minimum in the volume versus axial ratio plane starts to develop [78]. For $x \leq 0.9$, the second energy minimum becomes stable relative to the first one, and the system stabilizes in the ϵ-phase having a *hcp* structure with $c/a \leq 1.6$. The axial ratio in the ϵ-phase initially decreases with the Cu concentration, and above 20 At.-% Cu it shows an increase towards the ideal value of 1.63 [236]. It turned out that though the η- and ϵ-brasses have the same hexagonal crystal structure, they represent two different phases, each having its own local energy minimum [78].

8.3.2 Cubic Silver–Zinc Alloys

Today, very little is known about the elastic properties of binary metallic alloys for the entire range of concentrations in a given lattice structure. The solubility is usually limited to from a few percent up to 30–40%, and even in the thermodynamically stable range of concentrations the experimental information is meagre. On the other hand, results from Figures 8.11 and 8.12 clearly demonstrate that the EMTO method is suitable for an *ab initio* description of the equilibrium bulk properties as a function of concentration.

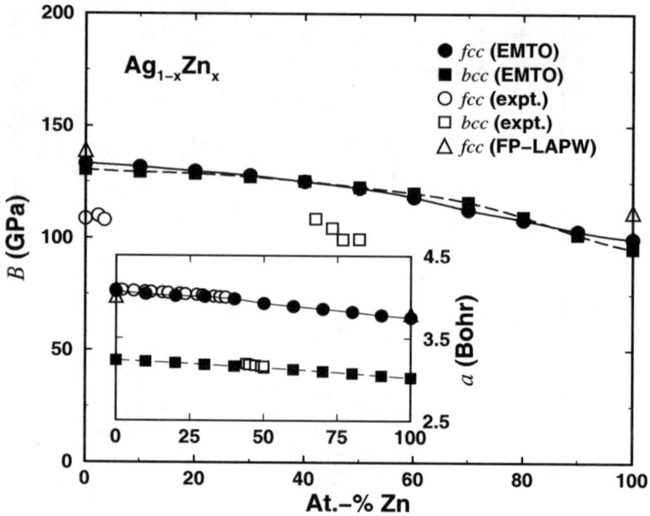

Fig. 8.13. Theoretical bulk moduli (B) and lattice constants (a, inset) of $Ag_{1-x}Zn_x$ alloys in the fcc and bcc structures, calculated using the EMTO method in combination with the LDA for the energy functional. The EMTO values are compared with the experimental data [232, 239]. The FP-LAPW results are shown for pure Ag [237] and Zn [238].

The EMTO method allows one to study both the thermodynamically stable and unstable phases of random alloys [115], and therefore it can be used to achieve a deep understanding of the dynamical stability and instability at different concentrations of alloying elements.

In Figure 8.13, we show the EMTO lattice parameters (a, inset) and bulk moduli (B) calculated for $Ag_{1-x}Zn_x$ alloys with $0 \leq x \leq 1$ in the fcc and bcc phases. For comparison, former theoretical results for the end-members [237, 238] and experimental data for the α- and β-phases [232, 239] are also shown. The theoretical results from this section were produced using the LDA. The agreement between the EMTO results for fcc Ag and Zn and those calculated using the LAPW method [237, 238] is good: the relative differences in a and B being 1.3% and −4.3% for Ag, and 0.3% and −10.9% for Zn, respectively. We note that for pure Ag the calculated volume per atom is 0.3% larger for the bcc lattice compared with that of the fcc lattice. For pure Zn the corresponding difference is 0.7%.

The theoretical and experimental elastic constants c' and c_{44} for fcc and bcc AgZn alloys are presented in Figures 8.14 and 8.15, respectively. The open symbols refer to experimental data from the Landolt–Börnstein tables [240] (α-phase) and Murakami and Kachi [239] (β-phase). For both elastic constants the present results are in excellent agreement with the experimental values. For pure Ag, the theoretical elastic parameters calculated using the FP-LAPW

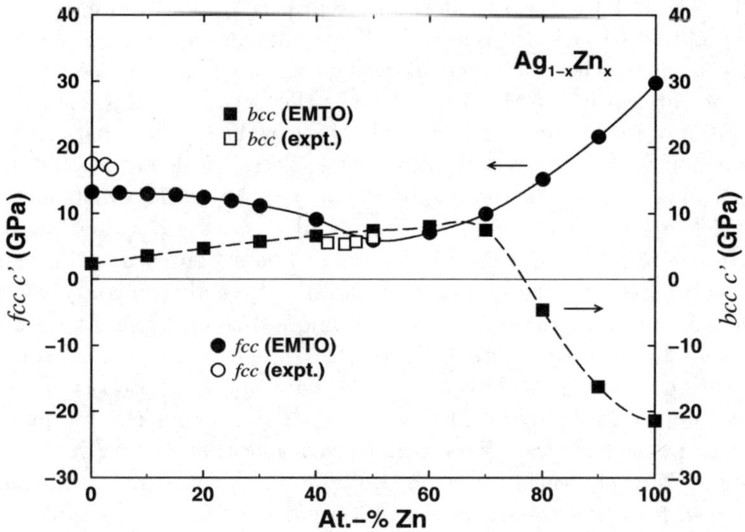

Fig. 8.14. Theoretical (EMTO-LDA) and experimental [239, 240] c' elastic constant for $Ag_{1-x}Zn_x$ alloys in the *fcc* and *bcc* structures [115].

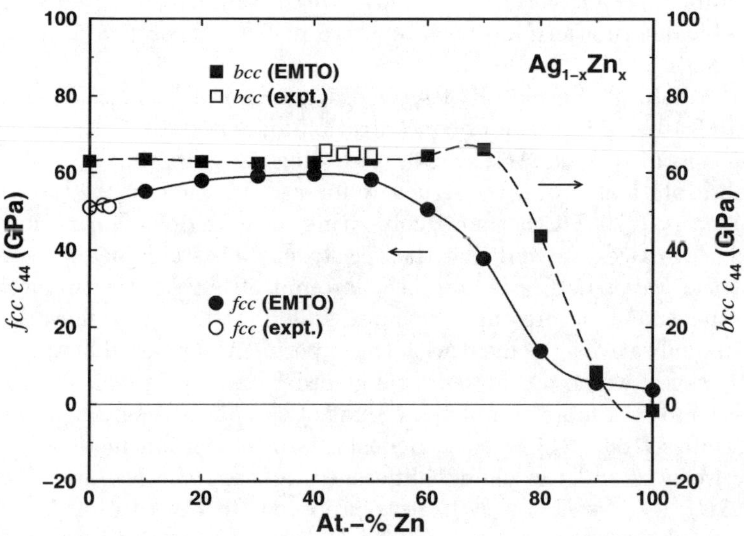

Fig. 8.15. Theoretical (EMTO-LDA) and experimental [239, 240] c_{44} elastic constant for $Ag_{1-x}Zn_x$ alloys in the *fcc* and *bcc* structures [115].

method [241] and FP-LMTO method [188] are $c'_{fcc} = 16$ and 21 GPa, and $c_{44,fcc} = 52$ and 61 GPa, respectively. If we take, as the error connected with such calculations, the difference between the two full-potential results, the agreement between the EMTO and former theoretical results is very good.

We point out that the calculated variations of the elastic constants with the concentration of Zn everywhere follow the experimental trends. In particular, the theoretical average softening of c' for *fcc* $Ag_{1-x}Zn_x$ with $x \leq 0.3$ is -0.07 GPa per At.-% Zn. The corresponding measured value lies between -0.07 and -0.22 GPa per At.-% Zn [240]. These figures may be compared to those obtained for the α-phase of CuZn, where theory and experiments give average softening of -0.30 GPa [78] and -0.23 GPa per At.-% Zn [189], respectively. Inspecting Figures 8.14 and 8.15, we find that except two narrow concentration intervals we have $c'_{fcc} > c'_{bcc}$ and $c_{44,fcc} < c_{44,bcc}$. This means that the Zener anisotropy (Equation (6.21)) is larger for the *bcc* phase than for the *fcc* phase, *i.e.* *fcc* AgZn is more isotropic than *bcc* AgZn.

In $Ag_{1-x}Zn_x$, the variation of the bulk modulus with the Zn concentration is smooth in both crystal structures (Figure 8.13). However, a quite different behavior is observed in the case of c' and c_{44}. For instance, c'_{bcc} is drastically reduced above $x \approx 0.78$, where, in fact, this system reaches a dynamical instability. At the same time, at large Zn content the *fcc* AgZn is strongly stabilized against tetragonal shear. On the other hand, c_{44} has almost constant value for $x \leq 0.5$ and $x \leq 0.7$ in the *fcc* and *bcc* structures, respectively. Above these concentration ranges c_{44} shows rapid variation in both structures, describing a trend from stable structures (Ag end) towards unstable (*bcc*) or barely stable (*fcc*) structures (Zn end). The instability of *fcc* Zn against orthorhombic deformations has been reported in other theoretical investigations as well [238].

The enthalpy of formation of $Ag_{1-x}Zn_x$ random alloy ($\Delta H(x)$) is plotted in Figure 8.16 as a function of concentration and crystal structure. The solid solution is formed when $\Delta H(x) < 0$. According to figure, the $Ag_{1-x}Zn_x$ alloys form solid solutions in the *fcc* structure for $x < 0.66$, and in the *bcc* structure for $0.19 < x < 0.70$. Furthermore, comparing the enthalpy of formation for the *fcc* and *bcc* lattices, we find that the *fcc* structure is stable below ~ 40 At.-% Zn and the *bcc* structure is stabilized around 50 At.-% Zn. In the Zn-rich region, the *hcp* structures are the most stable lattices (not shown). Figure 8.16 is in qualitative agreement with the experimental phase diagram of AgZn [234]. However, we note that short range order and entropy effects must also be considered in a more complete account of the phase diagram.

In studies [188, 243] of elastic constants of transition metals, a correlation between c' and the energy difference between the *bcc* and *fcc* structures, $\Delta E_{bcc-fcc} \equiv E_{bcc} - E_{fcc}$, was proposed. In the case of AgZn alloys $\Delta E_{bcc-fcc}(x)$ first decreases from 0.9 mRy ($x = 0$) to -0.4 mRy ($x = 0.6$), and then increases to 3.6 mRy ($x = 1$). This trend correlates reasonably well with both c'_{fcc} and $-c'_{bcc}$. However, a much better correlation exists between $\Delta E_{bcc-fcc}(x)$ and the difference between the two tetragonal shear moduli. At

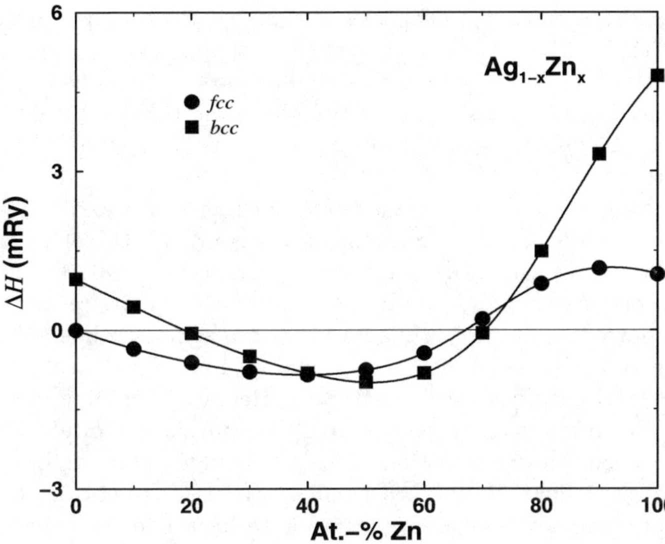

Fig. 8.16. Theoretical (EMTO-LDA) enthalpy of formation for $Ag_{1-x}Zn_x$ alloys in the *fcc* and *bcc* structures [115].

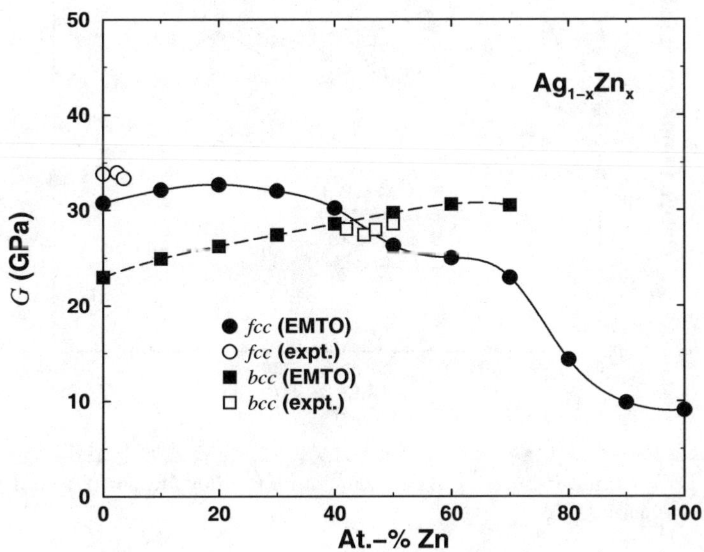

Fig. 8.17. Theoretical (EMTO-LDA) and experimental [239] shear modulus of $Ag_{1-x}Zn_x$ random alloys in the *fcc* and *bcc* crystallographic phases [115].

small Zn concentrations c'_{fcc} decreases and c'_{bcc} increases with x, suggesting the stabilization of the *bcc* phase against the *fcc* phase. In the region between 45 and 65 At.-% Zn the c'_{bcc} becomes slightly larger than the c'_{fcc}, showing the pronounced energy minimum for the *bcc* phase. Above 65 At.-% Zn one has $c'_{fcc} > c'_{bcc}$, and the re-stabilization of the *fcc* structure, relative to the *bcc* structure, occurs.

When a comparison can be made between experiments and EMTO results for the elastic constants, the agreement is very good. We therefore have confidence in our data also when they show an anomalous behavior. The abruptness of the variations in c' and c_{44} (Figures 8.14 and 8.15) is more pronounced than has been reported in any experiments for non-magnetic solids [240]. This is also seen in the polycrystalline shear modulus (Figure 8.17) calculated from the single crystal elastic constants using the Hershey averaging method (Section 6.3.1). The often-used "law-of-mixing" for the shear moduli, *i.e.* a linear variation between the values at the end members in the alloy, definitely fails in the case of AgZn alloys. Rapid, but smaller, variations of elastic constants as a function of composition or temperature have been noted for several transition metal alloys [240], and attributed to topological Fermi-surface transitions [242].

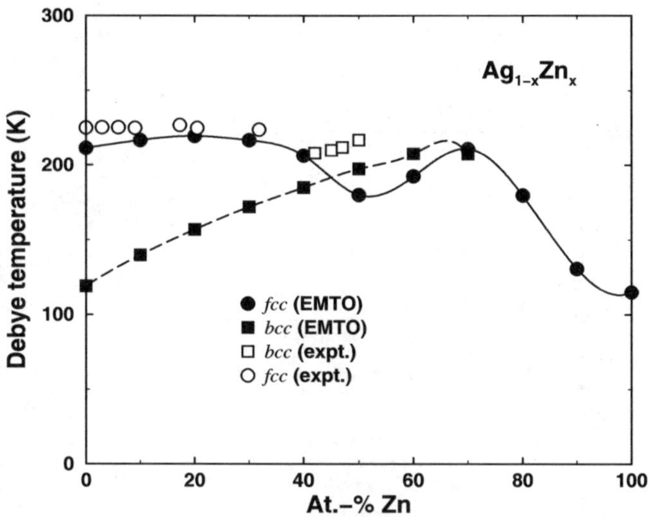

Fig. 8.18. The low temperature theoretical (EMTO-LDA) and experimental [239, 244] Debye temperatures of $Ag_{1-x}Zn_x$ random alloys in the *fcc* and *bcc* crystallographic phases [115].

Figure 8.18 presents the Debye temperatures for *fcc* and *bcc* AgZn calculated from Equations (6.27) and (6.26). Since the Debye temperatures are not defined for dynamically unstable lattices, for the *bcc* phase they are shown only for $x \lesssim 0.76$. For reference, the available low temperature experimental values are also included for the *fcc* [244] and *bcc* [239] structures. The agreement between experimental and theoretical Debye temperatures, including the concentration dependence, is very good. We recall that the quantity θ_D goes to zero as one approaches the critical concentration where the system becomes mechanically unstable [128]. In *bcc* AgZn, this happens near $x \approx 0.78$ (Figure 8.14). Note how narrow the region of rapid variation in θ_D is in Figure 8.18.

8.3.3 Hexagonal Silver–Zinc Alloys

The crystal structure gives the key to many properties of solid materials. Unusual structural properties that can be subtly tuned by chemical composition or external conditions are of great interest [89, 245, 246]. Most elemental metals crystallizing in the hexagonal close-packed (*hcp*) lattice have an axial ratio c/a that is close to the ideal value 1.633 obtained in a stacking of rigid spheres [236]. Zn-rich and Cd-rich alloys are exceptions, with $c/a > 1.75$.

The *hcp* structure is the thermodynamically most stable phase of $Ag_{1-x}Zn_x$ in two separate regions: in the ϵ-phase from about $x = 0.68$ to $x = 0.87$ and in the η-phase when $x > 0.95$. Experiments [232, 247, 248] show that c/a decreases very rapidly on alloying in the η-phase. In the ϵ-phase c/a is somewhat lower than the ideal value [247], first slowly decreasing as a function of Zn content, followed by a sudden and pronounced upturn close to the phase boundary. Contrasting this, the volume per atom varies slowly and monotonically in the *hcp* AgZn lattice (inset in Figure 8.13) with values for the intermediate ϵ-phase agreeing well with a simple interpolation from pure Ag to the Zn-rich η-phase. The striking variations in c/a on alloying in the AgZn solid solutions have been investigated using the EMTO method [235]. Since the GGA reproduces the equilibrium volume of pure Zn with higher accuracy compared to the LDA (see Section 8.3.1), it is also expected to lead to a more accurate hexagonal lattice constant than the LDA (Figure 8.12). Accordingly, all results from this section were obtained within the GGA [15].

To assess the accuracy of the EMTO method for the crystal structure and elastic constants of *hcp* random alloys, in Table 8.2 we compare the EMTO results obtained for the $Ag_{0.3}Zn_{0.7}$ random alloy with experimental data [248].

Table 8.2. Theoretical (EMTO-GGA) and experimental [248] equilibrium atomic radius w (in Bohr), hexagonal axial ratio $(c/a)_0$, and elastic constants (in GPa) of the *hcp* $Ag_{0.3}Zn_{0.7}$ random alloy.

	w	$(c/a)_0$	c_{11}	c_{12}	c_{13}	c_{33}	c_{44}
EMTO	2.98	1.579	110	56	63	129	27
expt.	2.92	1.582	130	65	64	158	41

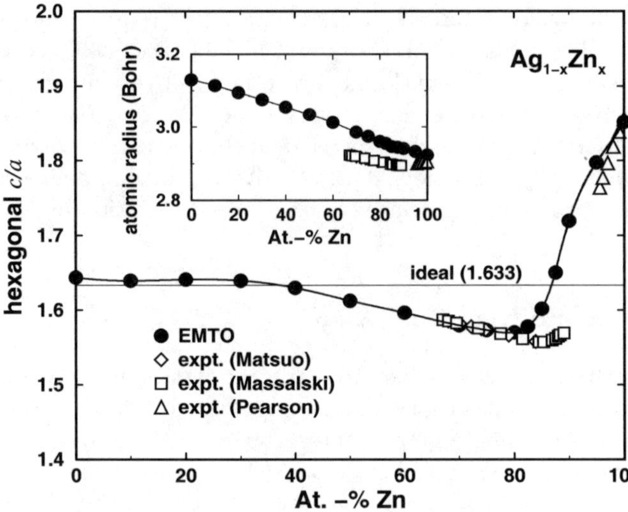

Fig. 8.19. Concentration dependence of the theoretical (EMTO-GGA) equilibrium axial ratio $(c/a)_0$ in *hcp* $Ag_{1-x}Zn_x$ alloys. The inset shows the calculated equilibrium atomic radii w as a function of Zn content. Experimental data are from Matsuo [248], Massalski [247] and Pearson [232].

The deviation between the theoretical and experimental equilibrium atomic radius and $(c/a)_0$ are 2% and 0.2%, respectively. The calculated elastic constants are somewhat small when compared with the measured values, but the relative magnitudes are well reproduced by the EMTO approach. The overall agreement between theory and experiment in Table 8.2 is very satisfactory, especially if one notes that the total energy minimum is very shallow in AgZn alloys, which makes the calculation of elastic properties numerically difficult.

Figure 8.19 shows the theoretical $(c/a)_0$ ratio for *hcp* $Ag_{1-x}Zn_x$ alloys in the whole range of concentrations $0 \leq x \leq 1$. Experimental data taken in the ϵ-phase [247, 248] and η-phase [232], are also included. Where a comparison with experiments is possible there is an excellent agreement between theory and experiment, which further testifies to the accuracy with which the EMTO approach can describe structural properties of AgZn random alloys. In contrast to the rapid changes in $(c/a)_0$, the equilibrium atomic radii w_0 follow, to a good approximation, Vegard's rule over the entire concentration range (insert in Figure 8.19).

In order to understand the conspicuous sharp upturn of $(c/a)_0$ near the upper concentration limit of the ϵ phase, Magyari-Köpe *et al.* [235] calculated the volume dependence of c/a for different concentrations. Figure 8.20 shows c/a as a function of the atomic radius w and chemical composition in the concentration range of interest. An important feature in the volume dependence of c/a is seen: for $x \leq 0.8$ c/a slightly decreases with volume, whereas

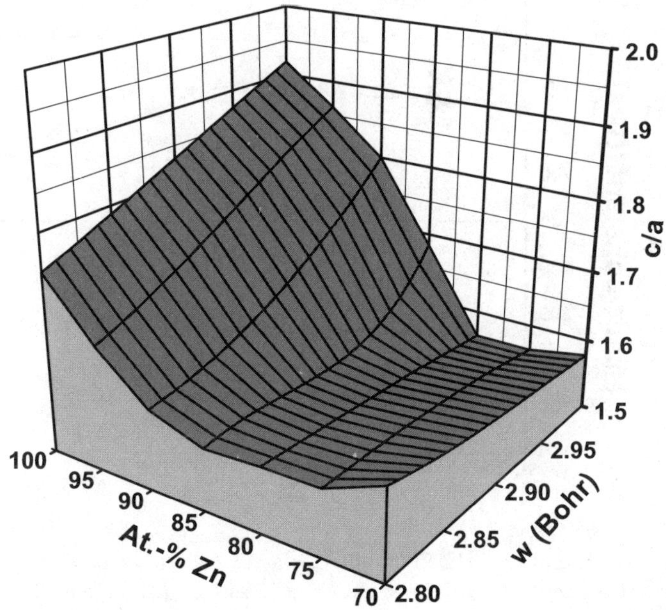

Fig. 8.20. Theoretical (EMTO-GGA) axial ratios (c/a) of *hcp* AgZn random alloys plotted against chemical composition (At.-% Zn) and average atomic radius (w).

for $x \geq 0.8$ it shows the behavior characteristic of pure Zn [249]. There is a narrow concentration range around 82% Zn where the volume dependence of c/a is nearly flat. The parameter R, describing the logarithmic volume derivative of c/a near the equilibrium structure (Equation (6.56)), has an almost constant value of -0.2 in the Ag-rich part of the ϵ-phase. However, at about 82% Zn there is a change in the sign of R, followed by a marked increase. This variation in R imposes a transition in the linear compressibility ratio K.[1] For $x \leq 0.82$ we have $K < 1$, *i.e.* the a axis is more compressible than the c axis. For these compositions K shows weak concentration dependence, which correlates well with the trend of the axial ratio from Figures 8.19 and 8.20. At concentrations above ~82% Zn, K becomes larger than 1 and increases rapidly with x. The hardening of a axis relative to c axis causes the upturn in the $(c/a)_0$ within the thermodynamic stability field of the ϵ phase.

The variation of the total energy $E(w, c/a)$ with c/a at a fixed volume, calculated around the equilibrium $(c/a)_0$, is described by the elastic constant c_s [132]. In the lower panel of Figure 8.21 the concentration dependence of the

[1] The ratio of the linear compressibilities parallel and perpendicular to the c axis is obtained as $K \equiv K_\parallel / K_\perp = (c_{11} + c_{12} - 2c_{13})/(c_{33} - c_{13})$, where c_{ij} are the hexagonal elastic constants.

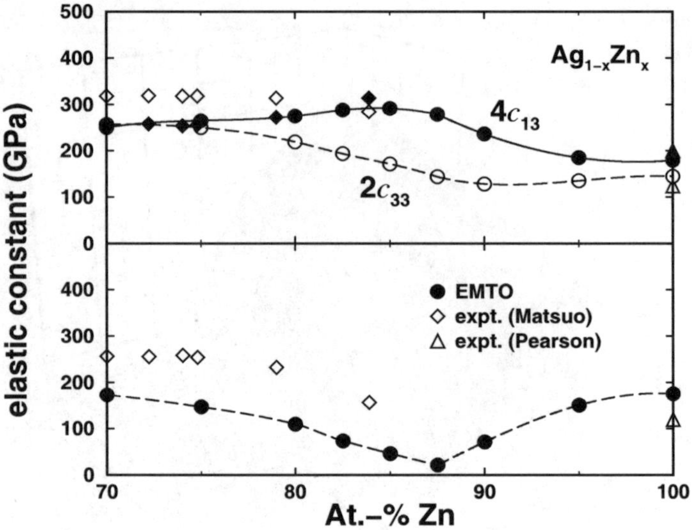

Fig. 8.21. Calculated (EMTO-GGA) elastic constants of *hcp* AgZn alloys as a function of concentration. Upper panel: $2c_{33}$ and $4c_{13}$. Lower panel: $c_s \equiv c_{11} + c_{12} + 2c_{33} - 4c_{13}$.

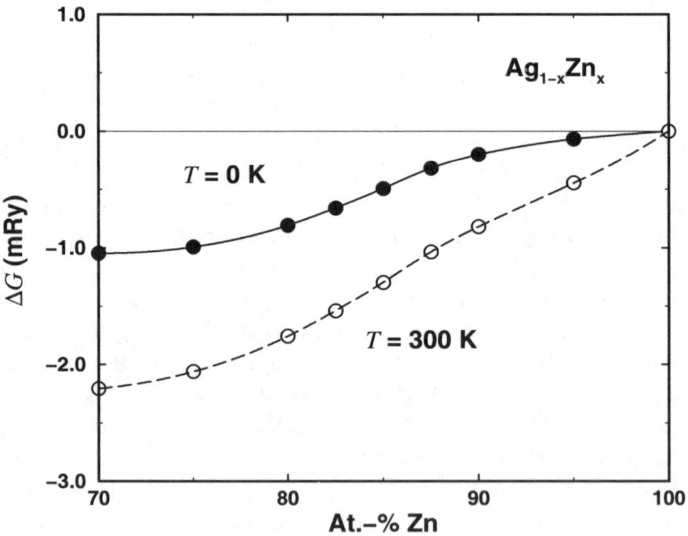

Fig. 8.22. Theoretical (EMTO-GGA) Gibbs energy of formation for *hcp* AgZn random alloys for temperatures of 0 (solid line) and 300 K (dashed line).

theoretical c_s is compared with the experimental data [232, 248]. Although there is an almost constant shift between theoretical and experimental values [248], the observed trend in $c_s(x)$ is well captured by the EMTO method. The pronounced minimum in $c_s(x)$ around 88% Zn appears as a result of the noticeable variations of $2c_{33}$ and $4c_{13}$ terms from the expression of c_s with concentration (see upper panel in Figure 8.21).

The calculated trend of the elastic constant $c_s(x)$ shows that the *hcp* AgZn random alloys may have a mechanical instability, or be very close to such a behavior, at about 88% Zn. Using our calculated total energies, we estimated the Gibbs energies of formation $\Delta G(x)$ of AgZn random alloys at $T = 0$ and $T = 300$ K.[2] This is shown in Figure 8.22. From the shape of $\Delta G(x)$ one can determine the stability limits of the ϵ and η phases by drawing a common tangent to the Gibbs energies calculated for these phases. We find that the theoretical two-phase-field region decreases from $0.77 \leq x \leq 1$ at $T = 0$ K to $0.83 \leq x \leq 0.96$ at $T = 300$ K, which is in qualitative agreement with the phase diagram information [234]. Thus, the softening of the hexagonal phases along the c axis, *i.e.* $c_s(x) \to 0$, will occur inside the two-phase-field region in the phase diagram, where in fact a single *hcp* phase is metastable and separates into ϵ and η phases.

Finally, we discuss the question of the rapid decrease of $(c/a)_0$ on adding Ag to η phase. Magyari-Köpe *et al.* [235] have shown that the anomalous $(c/a)_0$ ratio in Zn-rich η-phase has the same electronic origin as the one reported in the case of pure Zn [250, 251, 252]. According to that, in Zn the equilibrium $(c/a)_0$ ratio minimizes the band energy contribution to the total energy. With increasing Ag content, *i.e.* decreasing s electron density, the distortion-promoting band energy maintains its dominant role, and a reduced axial ratio minimizes the total energy.

8.4 Binary Transition-metal Alloys

In this section, we quote a few applications of the EMTO method to binary transition-metal alloys without a comprehensive description of the background and of the numerical calculations. For these details, the reader is referred to the corresponding references.

8.4.1 Iron–Magnesium Alloys at High Pressure

Iron and magnesium are almost immiscible at ambient pressure. The low solubility of Mg in Fe is due to very large size mismatch between the alloy

[2] Here the Gibbs energy of formation of $Ag_{1-x}Zn_x$ random alloy is approximated by $\Delta H - S_{\text{conf}}\Delta T$, where ΔH denotes the enthalpy of formation and S_{conf} is the configurational entropy estimated using the mean-field expression $-k_B[(1-x)\ln(1-x) + x\ln(x)]$.

components. However, the compressibility of Mg is much higher than that of Fe, and therefore the difference in atomic sizes between elements decreases dramatically with pressure. Dubrovinskaia *et al.* [253] have demonstrated in a series of experiments that high pressure promotes solubility of magnesium in iron. They have shown that at megabar pressure range more than 10 At.-% of Mg can dissolve in Fe which can then be quenched to ambient conditions.

In Figure 8.23, we show some EMTO results for pure $Fe_{0.95}Mg_{0.05}$ alloys. The calculations employed the GGA [15] for the exchange-correlation energy. The mixing enthalpy was calculated at 0 K, and the non-magnetic *hcp* phase of Fe and the *bcc* phase of Mg were chosen as standard states at all pressures. The reliability of the theoretical treatment is illustrated in the inset in Figure 8.23, where we compare the theoretical pressure–volume relations for *hcp* $Fe_{0.95}Mg_{0.05}$ alloy with the experimental result obtained for Fe-rich alloys containing 4.1 At.-% Mg. The excellent agreement between theory and experiment, seen in the figure, illuminates that theory correctly describes the FeMg solid solution.

As is seen in Figure 8.23, at low pressure the mixing enthalpy for FeMg alloys is large and positive, in agreement with the very low solubility of Mg in Fe. This is also in line with one of the well-known Hume–Rothery rule for metallic alloys, which states that the formation of disordered metallic alloys

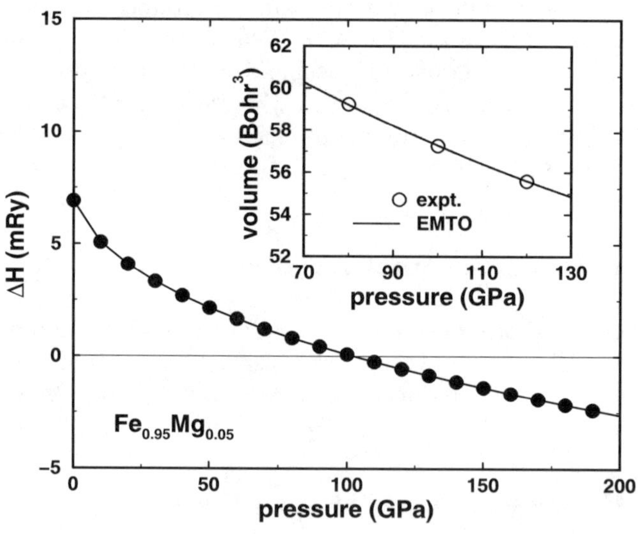

Fig. 8.23. Theoretical (EMTO-GGA) heat of formation for *hcp* $Fe_{0.95}Mg_{0.05}$ solid solution. The inset compares experimental (circles) and theoretical (solid line) pressure–volume dependence for *hcp*-structured FeMg alloy. The experimental data were obtained for Fe-rich alloys containing 4.1 At.-% Mg. $\Delta H < 0$ indicates a tendency of the system towards alloying and $\Delta H > 0$ represents a tendency towards phase segregation.

is very unlikely if atomic sizes of alloy constituents differ by more than 15% [254]. However, compressibility of Mg is much higher than that of Fe, and therefore the difference in atomic sizes between these two elements decreases dramatically with pressure. This raises the possibility that alloying of iron and magnesium may be more favorable under high-pressure conditions. Indeed, the EMTO calculations predict that the mixing enthalpy of disordered Fe-rich *hcp* FeMg alloys changes sign at about 100 GPa. This means that already at zero temperature there is a transition from the tendency towards phase separation between Fe and Mg at low pressure to the tendency towards alloying at higher pressure. With increasing temperature the tendency towards alloying increases further due to the entropy contribution.

8.4.2 Elastic Constants of Vanadium–Niobium Alloys

The vanadium-group transition metals (V,Nb and Ta) exhibit interesting and sometime anomalous properties. For example, they possess the largest electron–phonon interaction parameters among the transition metals [128] and Nb has the highest superconducting transition temperature among the elemental metals. Furthermore, in contrast to the normal decreasing behavior displayed by most metals, the elastic constants of these solids show irregular temperature dependence. Because of that, the V-group elements have been the subject of numerous experimental and theoretical investigations [255, 256, 257].

In Table 8.3, we compare the EMTO single-crystal shear moduli of *bcc* V and Nb with the available full-potential [188, 255] and experimental [189] data. The EMTO values were calculated using the GGA for the energy functional. The agreement between different sets of theoretical values is reasonable in the case of Nb, but large discrepancies can be seen for V. Part of these deviations have been ascribed [188] to the strong LDA over-binding in V, which is less pronounced for Nb (see Table 8.1). When compared to the experimental data, the FP-LMTO calculation using the GGA functional [255] is found to give slightly more accurate elastic constants than the EMTO method. In this

Table 8.3. Theoretical and experimental elastic constants of *bcc* V and Nb. The EMTO calculations were based on the GGA for the energy functional. The experimental values are given for room temperature and 0 K (numbers in parenthesis) [189, 258].

	Vanadium			Niobium		
	EMTO	FP-LMTO	expt.	EMTO	FP-LMTO	expt.
c'	72.6	67.3^a,37.5^b	54.5(57)	68.8	63.9^a,60^b	56.0(60)
c_{44}	29.7	37.4^a,5^b	42.6(46)	35.3	25.5^a,27^b	28.7(31)

^{a} GGA [255], ^{b} LDA [188].

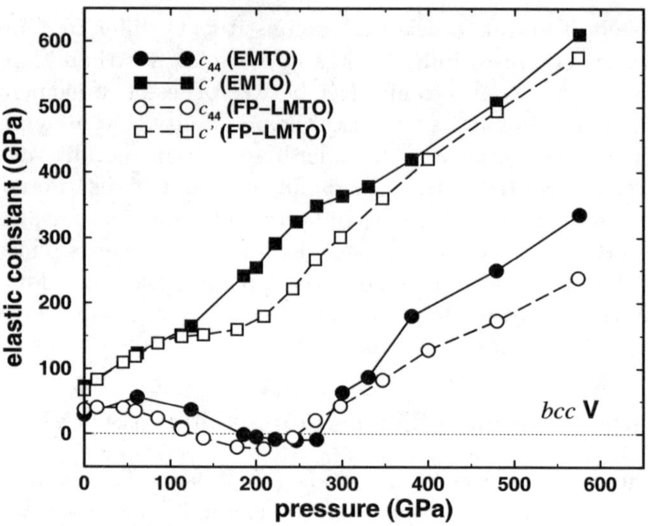

Fig. 8.24. Calculated pressure dependence of the shear elastic moduli of *bcc* V [255].

comparison, however, one should also take into account that both c' and c_{44} of V and Nb show strong anomalous temperature dependence below the room temperature. The average differences between the EMTO and experimental shear moduli for V and Nb are 32% and 23% at room temperature and 31%

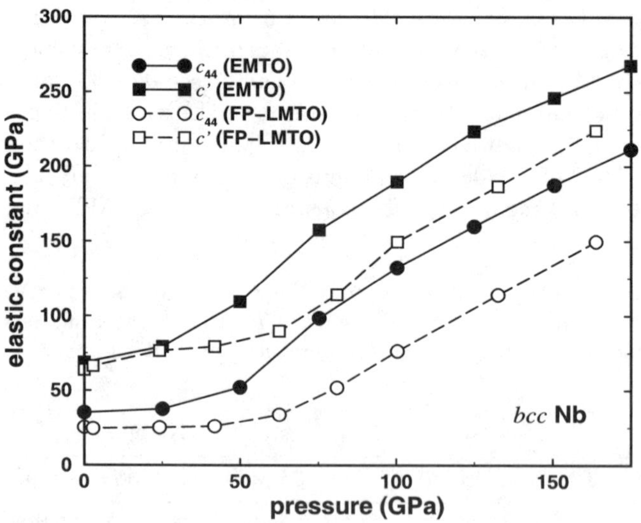

Fig. 8.25. Calculated pressure dependence of the shear elastic moduli of *bcc* Nb [255].

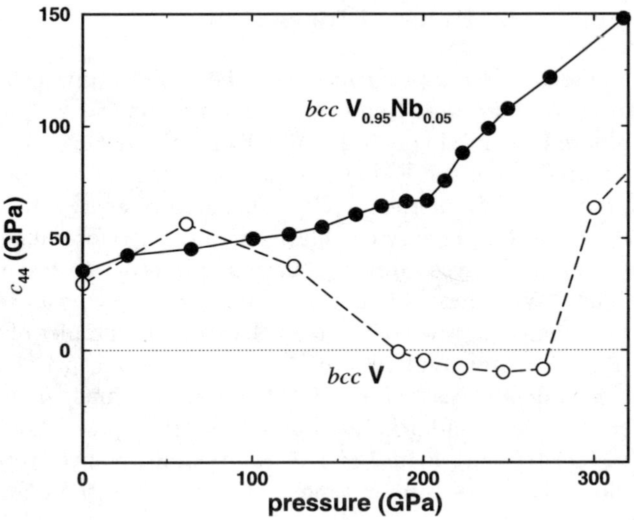

Fig. 8.26. Calculated (EMTO) pressure dependence of the c_{44} shear elastic modulus of *bcc* $V_{0.95}Nb_{0.5}$ random alloy [255]. For comparison, the EMTO results for pure V are also shown.

and 14% at 0 K, respectively. These are reasonable errors [188], especially if one takes into account that the theoretical elastic constants depend sensitively on the details of the calculations.

Figures 8.24 and 8.25 show the calculated pressure dependence of the shear elastic moduli of *bcc* V and Nb, respectively. There is a good parallelism between the EMTO and FP-LMTO [255] data. In particular, both methods reveal a mechanical instability in c_{44} for V. The corresponding pressures are slightly different in EMTO (180–270 GPa) and FP-LMTO (150–250 GPa) calculations. For Nb, both methods give softening of c' and c_{44} at pressures $\sim$ 50 GPa. Except these critical pressures ranges, the tetragonal shear modulus for V and Nb and c_{44} for Nb exhibit normal increasing behavior with pressure. The theoretically predicted structural instability in *bcc* V is in line with the experimentally observed structural phase transition at 69 GPa [257].

Landa *et al.* [255] proposed that the above pressure-induced shear instability (softening) of V (Nb) originates from the peculiar electronic structure. They have shown that substitution of 5 At.-% of V with Nb removes the instability of V with respect to trigonal distortions in the vicinity of 200 GPa pressure, but still leaves the softening of c_{44} in this pressure region. The pressure dependence of c_{44} for *bcc* $V_{0.95}Nb_{0.5}$ random alloy, calculated using the EMTO method in combination with GGA, is shown in Figure 8.26. We can see that, apart from significant softening around $\sim$ 200 GPa, in Nb-bearing V the stability criteria for a cubic crystal (Section 6.2) holds within the whole pressure range considered.

8.4.3 Ferromagnetic Fe-based Binary Alloys

In order to demonstrate the performance of the EMTO approach for calculating the total energy of ordered and random ferromagnetic alloys, Kissavos *et al.* [259] compared the mixing energies of ferromagnetic FeCo, FeNi and FeCu systems calculated using the EMTO method and the full-potential projected augmented wave (PAW) method [169]. The mixing energy reflects directly the accuracy of the total energy calculations, and, therefore, it is an excellent quantity to compare between different methods in order to test their relative accuracies. The PAW is generally known to be among the most accurate methods for the electron structure calculations. The mixing energies of the Fe-based systems in B2, L10, and random *fcc* phases are plotted in Figure 8.27. The mixing energy is defined as the energy of the alloy minus the concentration average energy of the standard states. For the latter, pure alloy components in the *bcc* or *fcc* structure for B2 or L10 compounds, respectively, were used. For the random *fcc* alloys (upper panel in Figure 8.27), the reference state is again *fcc*.

In EMTO calculations, the CPA is used to model the chemical disorder, whereas the PAW modeling is based on the special quasirandom structures method [67, 259]. As can be seen in Figure 8.27, PAW and EMTO results fall almost on top of each other, showing that the EMTO method can be used with confidence for calculations of total energies for chemically ordered, as well as disordered alloys.

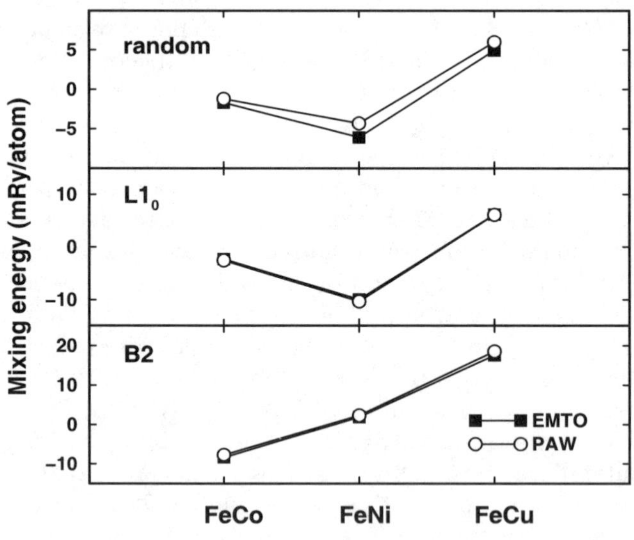

Fig. 8.27. Mixing energies for the B2 (lower panel), $L1_0$ (middle panel), and random *fcc* (upper panel) phases in FeCo, FeNi, and FeCu systems, calculated with both the EMTO and the PAW method [169].

8.4.4 Paramagnetic Fe-based Binary Alloys

Paramagnetic alloys are often treated as nonmagnetic systems in *ab initio* calculations. Accordingly, the local magnetic moments are artificially suppressed on alloy components. However, for a majority of the ferromagnetic alloys the local moments survive above the Curie temperature, though the net magnetization indeed vanishes due to their random orientations. Iron-based alloys are typical examples of such systems. On the other hand, the *disordered local moment* (DLM) approach [260, 261] allows one to model the effect of the totally random magnetic moments. In practice, a binary $A_{1-x}B_x$ alloy with a complete disorder of local magnetic moments on site A is simulated as a three-component alloy $A^{\uparrow}_{(1-x)/2}A^{\downarrow}_{(1-x)/2}B_x$, where arrows represent different spins. Similarly, a four-component alloy $A^{\uparrow}_{(1-x)/2}A^{\downarrow}_{(1-x)/2}B^{\uparrow}_{x/2}B^{\downarrow}_{x/2}$ is considered when magnetic disorder is present on both sites. Treating a three- or four-component alloy instead of a binary alloy is an easy task within the CPA, but it is practically impossible with any alternative approaches.

The EMTO mixing enthalpies for paramagnetic *fcc* FeCo, and *bcc* FeCr and FeV alloys calculated using the GGA energy functional [259] are plotted in Figures 8.28–8.30. For these systems experimental mixing enthalpies of the paramagnetic phases are available in the whole concentration range [262]. The calculated results for the *fcc* FeCo alloys agree very well with the experimental values, especially considering the very small absolute values of mixing enthalpy of random *fcc* FeCo alloys. For *bcc* FeCr and FeV alloys,

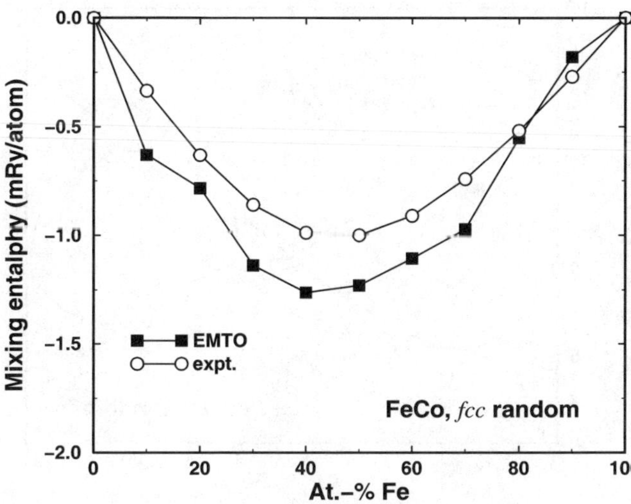

Fig. 8.28. Mixing enthalpies for paramagnetic random *fcc* FeCo alloys calculated using the EMTO method in combination with the DLM model and the GGA energy functional (squares) and the corresponding experimental values (circles) [262].

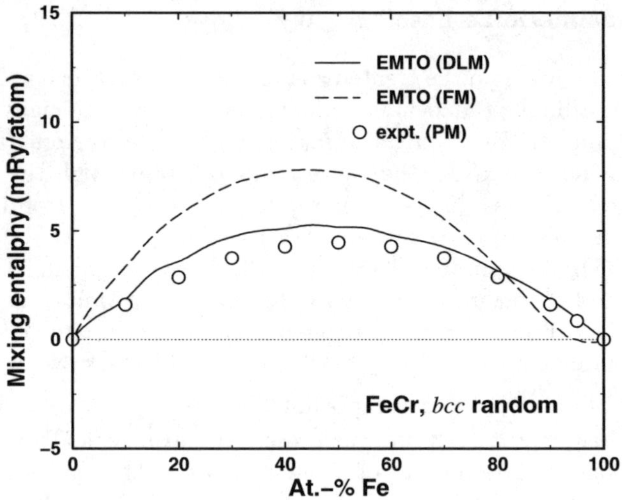

Fig. 8.29. Mixing enthalpies for paramagnetic (solid line) and ferromagnetic (dashed line) random *bcc* FeCr alloys calculated using the EMTO method in combination with the DLM model and the GGA energy functional and the corresponding experimental values (circles) for the paramagnetic *bcc* FeCr alloys [262].

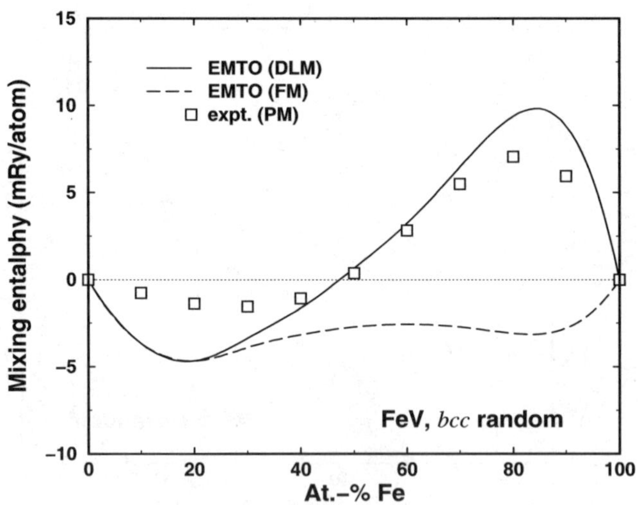

Fig. 8.30. Mixing enthalpies for paramagnetic (solid line) and ferromagnetic (dashed line) random *bcc* FeV alloys calculated using the EMTO method in combination with the DLM model and the GGA energy functional and the corresponding experimental values (squares) for the paramagnetic *bcc* FeV alloys [262].

again, one can see good agreement between the corresponding data sets. For FeCr, both the EMTO-DLM results and the experimental data measured for the paramagnetic phase show nearly parabolic dependence on Fe concentration, following the regular solid solution model. On the other hand, for FeV alloys there are strong deviations from the regular sold solution behavior, with a spectacular change of sign of the mixing enthalpy around the equiatomic composition. The EMTO method can capture this remarkable feature.

To demonstrate the importance of magnetic disorder in FeCr and FeV systems, in Figures 8.29 and 8.30 the mixing enthalpies calculated for the ferromagnetic phases [259, 264] are also shown (dashed lines). We observe substantial deviations between the ferromagnetic and the DLM results. In particular, FeV alloys in the ferromagnetic phase have negative mixing enthalpy within the entire concentration range, in contrast to the paramagnetic phase, where the Fe-rich alloys are not stable. Different signs of the mixing enthalpies in different magnetic phases can also be seen for the dilute Fe-rich FeCr alloys. According to the EMTO results from Figure 8.29, at low Cr concentrations the ferromagnetic FeCr alloys are anomalously stable. This prominent finding was confirmed by full-potential calculations based on the PAW method combined with special quasirandom structures [265] as well as with the supercell technique [266]. Furthermore, it has been suggested that the anomalous mixing between Fe and Cr below $\sim$ 10 At.-% Cr has an electronic–magnetic origin [265, 267].

8.5 Cerium–Thorium Alloys

The structural stability of Ce, Th, and the CeTh system as a function of compression has been investigated using the EMTO method [268]. At low pressure, Ce adopts the *fcc* crystal structure, which transforms to orthorhombic or body centered monoclinic below 10 GPa. Above 12 GPa, Ce is stabilized in a body-centered tetragonal (*bct*) structure. There are two iso-structural *fcc* phases of Ce, namely γ-Ce and α-Ce. The $\gamma \rightarrow \alpha$ transition occurs close to 1 GPa and is associated with a Mott transition of the f electrons from a localized (γ-Ce) to a delocalized (α-Ce) state. Thorium crystalizes in *fcc* structure at low pressures (α-Th), and transforms to *bct* at about 60 GPa. At megabar pressures, both metals remain in a *bct* crystal [269, 270, 271] with a c/a axial ratio close to 1.65. The crystal structure of the CeTh system shows a similar behavior as a function of pressure [272, 273].

The theoretical equilibrium properties of α-Ce and α-Th are compared with the experimental data in Table 8.4. The theoretical equations of state were computed using the GGA for the exchange-correlation energy. We notice that EMTO calculations give excellent equilibrium properties of both Ce and Th. The somewhat larger deviation between EMTO and experimental bulk properties of CeTh alloy can be understood if one takes into account that the

Table 8.4. Theoretical and experimental equilibrium atomic volume (V_0 in Bohr^3) and bulk modulus (B_0 in GPa) for Ce, Th and $Ce_{0.47}Th_{0.53}$. Theoretical results were obtained using the EMTO method and the FP-LMTO method [274]. For Ce, equilibrium volumes are given for both α and γ (parenthesis) phases.

	Ce			Th			$Ce_{0.47}Th_{0.53}$	
	EMTO	FP-LMTO	expt.	EMTO	FP-LMTO	expt.	EMTO	expt.
V_0	187.1	176.3	$189.1^a(232.2^b)$	224.9	200.0	222.2^c	212.1	222.6^d
B_0	38	49	29^a	58	63	58^c	46	28^d

^{a} expt. α–Ce [275], ^{c} expt. α–Th [276],
^{b} expt. γ–Ce [277], ^{d} expt. [272, 273].

CeTh system is composed of γ-Ce and α-Th, whereas the EMTO calculations were performed on an alloy created by α-Ce and α-Th [268].

The calculated tetragonal c/a axial ratios for *bct* Ce and Th together with experimental data are shown in Figure 8.31. The *bct* structure is generated by distorting the *fcc* lattice along the Bain path [278], so that $c/a = \sqrt{2}$ corresponds to the undistorted *fcc* structure. Restricting the crystal structure

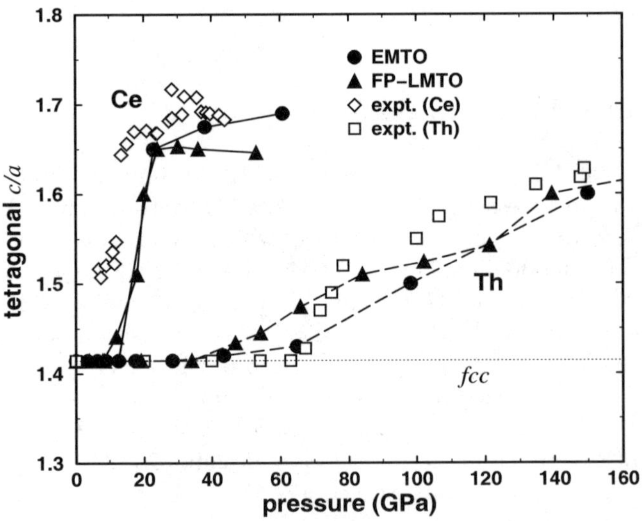

Fig. 8.31. The tetragonal c/a axial ratio for the *bct* structure of Ce and Th as a function of pressure. Experimental data for Ce [269] are marked with diamonds, and for Th [270, 271] with squares. The theoretical results are connected by solid lines for Ce and dashed lines for Th. The FP-LMTO results [274] are shown by triangles, and the EMTO results by circles. Horizontal line marks the c/a corresponding to the *fcc* lattice.

optimization to the *bct* symmetry, from the EMTO calculations (shown by circles) we find that Ce remains in high-symmetry *fcc* lattice up to ∼ 12 GPa, where a steep *fcc* → *bct* transition can be monitored. The *bct* phase of Ce with c/a =1.65−1.68 remains stable up to at least 60 GPa. These findings are in quantitative agreement with the experimental [269] and also with former FP-LMTO [274] results.

A somewhat different behavior is found for Th (Figure 8.31). According to the EMTO results, Th remains stable in its ambient pressure *fcc* phase up to ∼ 60 GPa. At higher compression, it transforms continuously into the *bct* phase. Note that the transition pressure is considerably higher in Th than in Ce. As in the case of Ce, the EMTO results for Th agree well with those of previous FP-LMTO calculations [274] and experimental data [270, 271], also shown in Figure 8.31.

The calculated and measured c/a axial ratio as a function of pressure for the $Ce_{0.43}Th_{0.57}$ disordered alloy is shown in Figure 8.32. The EMTO calculations confirm that the *fcc* → *bct* phase transition occurs between 10 and 20 GPa, which is close to the corresponding transition in Ce, but considerably lower than for Th. The fact that the transition pressure is a strongly nonlinear function of Th concentration in the CeTh system is in line with experimental observation [272, 273].

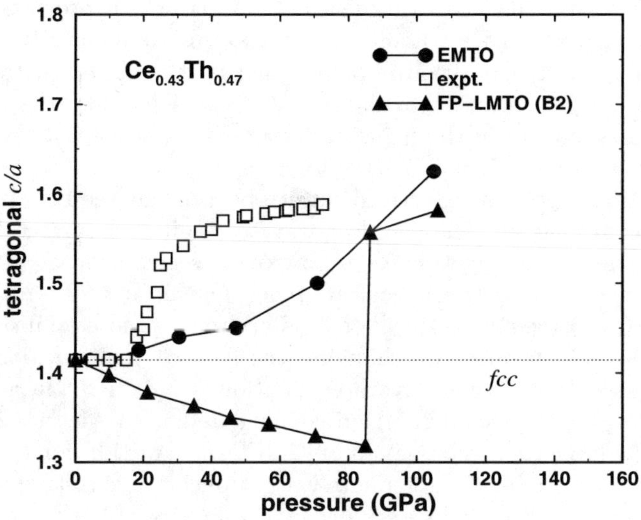

Fig. 8.32. The tetragonal c/a axial ratio for the *bct* structure of the $Ce_{0.43}Th_{0.57}$ disordered alloy as a function of pressure. Experimental data [272, 273] are marked with squares. The EMTO results are given by filled circles, and the results of FP-LMTO calculations for ordered CeTh compound with B2 structure [279] are given by triangles. The horizontal line marks the c/a corresponding to the *fcc* lattice.

The structural behavior of CeTh alloy was previously modeled by an ordered (B2) CeTh compound using the FP-LMTO method [279]. These earlier calculations predicted that the tetragonal axial ratio first decreases with pressure and then suddenly jumps to a high value closer to the measure value at a higher compression (shown by triangles in Figure 8.32). It was suggested that the discrepancy with experiment was due to the failure of modeling the disordered alloy with an ordered compound. The EMTO results from Figure 8.32 validate this assumption, and demonstrate that the disorder needs to be properly accounted for to accurately describe $Ce_{0.43}Th_{0.57}$.

8.6 Surface Concentration Profile

Surfaces form the fundamental interface for many physical and chemical interactions. In alloys, the surface chemistry may show significant alloying and temperature dependence. One important example is the threshold behavior in stainless steels [280, 281]. It is a well known fact that 13% or more Cr renders FeCr alloys an excellent corrosion resistance, but below this threshold the system behaves much like pure iron. Several phenomenological models tried to describe this compositional threshold behavior. According to a thermodynamic-kinetic passivation mechanism [282, 283], a stable Cr-rich oxide layer is formed on the surface of FeCr alloy, which protects the system against various attacks in different chemical environments. It is clear that the surface oxide composition is determined by the actual metal concentration in the surface prior to oxidation. Motivated by this, the chemistry of FeCr surfaces have been the focus of numerous experimental and theoretical investigations [284, 285, 285, 287].

Promoting or inhibiting certain surface phenomena requires full control of the surface properties. The simplest way to produce a surface having optimal properties for a specific task is to exploit segregation at alloy surfaces. The most widely used theoretical approach for surface segregation is based on Monte-Carlo technique [288]. In this approach, the equilibrium segregation profile is determined by changing the surface configuration repeatedly until energetically the most favorable solution is obtained. Because the total energy of a particular configuration is calculated from effective interactions, a Monte-Carlo method offers a simple and fast algorithm for multilayer surface segregation studies. Unfortunately, its application is restricted to systems with an ideal undistorted underlying lattice and where reliable effective interaction parameters can be constructed. Based on the EMTO method, Ropo *et al.* [289, 290, 291] introduced an alternative technique, which is suitable to determine the surface concentration profile in multi-component random alloys with arbitrary crystal structure. This approach treats the surface system as a grand canonical ensemble, allowing for an unconditional particle exchange between the surface region and a bulk reservoir.

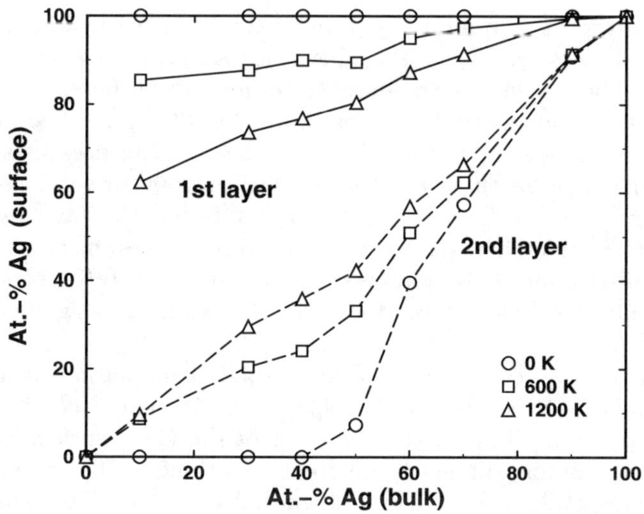

Fig. 8.33. Theoretical segregation profile calculated using the EMTO-LDA method for the (111) surface of AgPd as a function of temperature and bulk Ag concentration. Symbols connected by solid lines represent the top surface layer concentration of Ag, while symbols connected by dashed lines correspond to Ag concentration in the subsurface layer. Solid lines are intentionally ended at the Pd-rich end, in order not to give an incorrect impression about the segregation profile in dilute alloys.

The palladium–silver system represents a perfect case to illustrate the surface concentration profile calculations, since these alloys have continuous solid solubility in the *fcc* crystallographic structure. Figure 8.33 shows the top layer and the subsurface layer concentrations for the (111) surface of $Pd_{1-x}Ag_x$ alloys calculated as a function of bulk concentration (x) and temperature using the EMTO method and the LDA energy functional. The surface concentration profile was obtained by minimizing the grand potential of the surface subsystem. The condition for the minimum was expressed as the equilibrium between the surface and bulk effective chemical potentials (ECP). Here, the ECP is the difference between the Ag and Pd chemical potentials and is calculated from the energy change when a Pd atom was exchanged with Ag. At $T = 0$ K, the minimum requirement for the grand potential leads to the condition that the difference between the effective chemical potentials for the surface and bulk subsystems should be equal to the equilibrium segregation energy [290]. The latter vanishes for an equilibrium concentration profile where all the alloy components have nonzero concentration. For calculations at $T > 0$ K, the entropy was approximated by the configurational entropy. For more details about the thermodynamical model for metal surfaces, the reader is referred to the works by Ropo *et al.* [289, 290, 291].

Figure 8.33 indicates that at 0 K the first surface layer is completely filled with Ag for the whole bulk concentration interval. In the second layer, the

Ag concentration is zero for $x \leq 0.4$, and rapidly increases with increasing x for $x > 0.4$. In Ag-rich alloys, the subsurface Ag layer concentration is close to $(2x - 1)$. The stability of the pure Ag terminated surfaces at 0 K is due to the lower surface energy of Ag compared to that of Pd (see Section 7.1.3).

With increasing temperature, the Ag concentration decreases in the top layer and increases in the second layer. The entropy driven changes in the second layer are small for $x < 0.2$ and $x > 0.6$, but they are important for intermediate x values. The top layer concentration shows the largest variations with T in Pd-rich alloys. It is interesting that at $T \geq 600$ K, the second layer is almost bulk-like, whereas the top layer still contains a significant amount ($> 60\%$) of Ag.

In Table 8.5, we compare the EMTO equilibrium surface concentrations with the available experimental data [292, 293]. Wouda *et al.* [292] measured the chemical composition in the top layer of the (111) surface of $Ag_{33}Pd_{67}$ alloy using a scanning tunnelling microscopy technique. The reported surface concentrations of Ag are somewhat higher than the EMTO values. Reniers [293] investigated the surface composition of $Ag_{50}Pd_{50}$ alloy using Auger electron spectroscopy (AES). In these experiments the surface composition was estimated from the Auger current. Due to the approximate relation between current and concentration, there were significant uncertainties associated with

Table 8.5. Theoretical (EMTO-LDA) and experimental surface concentration profile for AgPd alloys as a function of bulk Ag concentration and temperature (T). The three concentrations refer to the Ag concentration in the bulk and in the 1st and 2nd surface layers, respectively.

method	bulk	1st layer	2nd layer	T (K)
EMTO[a]	30	88	21	600
"	30	80	29	900
expt.[b]	33	95	-	820
"	<33	93	-	720
"	<33	91	-	770
"	<33	91	-	820
"	<33	90	-	920
EMTO[a]	50	90	34	600
"	50	85	39	900
expt.[c]	50	70-99	-	673–873
expt.[d]	50	54-63	-	673–873

[a] [290], [b] [292],
[c] AES data without matrix correction [293],
[d] AES data with matrix correction [293].

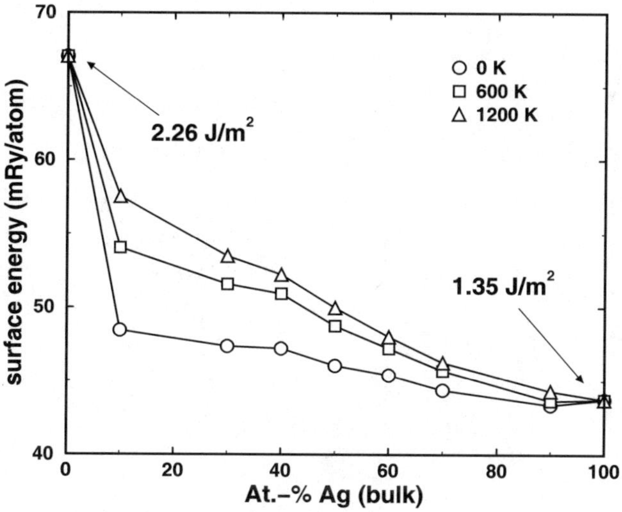

Fig. 8.34. Surface energy for the (111) surface of AgPd calculated using the EMTO-LDA scheme as a function of temperature and bulk Ag concentration. For pure metals the surface energies per unit surface area are also shown.

these measurements. Within these uncertainties the EMTO theoretical values are in reasonable agreement with the AES measurements.

The surface energy $\gamma(T, x)$ for the (111) surface of $Pd_{1-x}Ag_x$ alloys is plotted in Figure 8.34 as a function of bulk composition and temperature. The EMTO surface energies for pure Ag and Pd are 43 and 67 mRy/atom when expressed per surface atom, and 1.35 and 2.26 J/m^2 when expressed per surface area. These values are significantly larger than 0.96 and 1.64 J/m^2 listed in Table 7.4. The $\sim 28\%$ average difference is due to the gradient correction in the GGA exchange-correlation functional compared to the LDA.

At 0 K, the surface energy is mainly determined by the pure Ag surface layer, which is reflected by an almost flat $\gamma(0\text{K}, x) \approx \gamma_{\text{Ag}}$ line for $x \gtrsim 0.1$. With increasing temperature, $\gamma(T, x)$ converges towards the value estimated using a linear interpolation between end members. Note that the temperature dependence of γ is very similar to that of the surface Ag concentration from Figure 8.33. Although, at intermediate concentrations, the concentration from the subsurface layer shows strong temperature dependence, this effect is imperceptible in the surface energy. Therefore, the variation of the surface energy of an alloy with temperature and bulk composition is, to a large extent, governed by the surface layer, and the subsurface layers play only a secondary role.

9

Iron–chromium–nickel Alloys

In Chapters 7 and 8, we demonstrated the application of the EMTO method to ordered systems as well as to disordered binary alloys. In this chapter, we shall illustrate this in the case of FeCrNi-based alloys, which form the basis of austenitic stainless steels. First, the atomistic model of these important class of materials will be defined. In Section 9.2, we shall present the theoretical calculation of the elastic constants of FeCrNi alloys. After establishing the accuracy of the EMTO approach for this problem, we shall display the elastic property–chemical composition maps. The computed elastic properties will be used to determine the misfit parameters for quaternary alloying additions M(=Al,Si,V,Cu,Nb,Mo,Re,Os and Ir). In Section 9.3, the *ab initio* determination of the stacking fault energy of FeCrNi ternary alloys will be introduced. In connection with the stacking faults, we shall discuss the role of magnetic fluctuations on the mechanical properties of austenitic steels. A few important numerical details of the EMTO calculations presented in this chapter are listed in Section C.3.

9.1 Modeling the Alloy Steels

In addition to Fe and carbon, alloy steels contain several other elements in order to obtain specific properties of the material. In particular, *stainless steels* are alloy steels containing more than 13% Cr, which makes these alloys resistant against corrosion in various chemical environments [280, 281]. Stainless steels dominate both industrial and everyday applications of steels, where a combination of good corrosion resistance with high strength, stiffness, and toughness is required.

Austenitic stainless steels, the largest sub-category of stainless steels, contain a significant amount of substitutional Ni. The presence of Ni changes the crystal structure of steel at ambient temperatures from the *bcc* structure of α-Fe (ferrite) to the *fcc* structure of γ-Fe (austenite) [294]. The FeCrNi austenite possesses a rather unique combination of physical, mechanical, and chemical

Table 9.1. Some typical commercial stainless steel compositions from the AISI 300 series. Concentrations are given in atomic as well as in weight (parenthesis) percent.

AISI	C	Mn	P	S	Si	Cr	Ni	Mo	Fe
304	0.4	2	0.08	0.05	1.9	19–21	7–9	–	balance
	(0.08)	(2)	(0.045)	(0.03)	(1.0)	(18–20)	(8–10)	–	(balance)
310	1.1	2	0.08	0.05	2.9	25–27	17–20	–	balance
	(0.25)	(2)	(0.045)	(0.03)	(1.5)	(24–26)	(19–22)	–	(balance)
316	0.4	2	0.08	0.05	1.9	17–19	9–13	1–2	balance
	(0.08)	(2)	(0.045)	(0.03)	(1.0)	(16–18)	(10–14)	(2–3)	(balance)
317	0.4	2	0.08	0.05	1.9	19–21	10–14	2	balance
	(0.08)	(2)	(0.045)	(0.03)	(1.0)	(18–20)	(11–15)	(3–4)	(balance)

properties that make austenitic stainless steels suitable for many specific applications. The most common austenitic grades, belong to the AISI (American Iron and Steel Institute) 300 series. Some typical commercial stainless steel compositions from this series are listed in Table 9.1.

The amount of interstitial carbon in austenitic stainless steels is usually below 0.4 At.-%, but in some cases it could be as high as ~ 1 At.-%. If Cr-rich carbides are formed (usually, at grain boundaries), the alloy matrix becomes depleted of Cr and thereby its corrosion resistance is drastically reduced. This causes various forms of localized corrosion attack, such as intergranular corrosion, pitting corrosion, or stress-corrosion cracking. In order to prevent the formation of Cr-rich carbides one needs to keep the C content in alloy steels at low levels. This is achieved by additional alloying with strong carbide-forming elements like Mo, V, Nb, *etc.* The amount of impurities (*e.g.*, S or P), which may cause similar problems with the resistance of steel to localized forms of corrosion, is also kept at the lowest possible level [295].

The beneficial properties of austenitic stainless steels derive from the properties of the main phase in these alloys. According to Table 9.1, to a good approximation the latter can be considered as a solid solution formed mainly by Fe, Cr and Ni. This is a simplified atomistic model for alloy steels, where the different species are randomly distributed on an *fcc* underlying lattice and the interstitials are completely omitted.

The most common austenitic stainless steels have very low magnetic permeability. They show almost no response to a magnet and hence are generally regarded as non-magnetic metals. On the other hand, the main constituent transition metals are magnetic and have persistent disordered magnetic moments in their high temperature paramagnetic states [296, 297, 298, 299, 300, 301, 261, 302, 303, 304]. Actually, magnetic studies indicate that FeCrNi alloys also exhibit a rich variety of magnetic phases. At low temperatures, their magnetic structure is antiferromagnetic, spin-glass or ferromagnetic, depending on the Ni content [294]. The susceptibility of $Fe_{0.8-n}Cr_{0.2}Ni_n$ alloys with

$0.14 < n < 0.21$ was obtained to follow the Curie–Weiss law at temperatures above 26–130 K with effective magnetic moments between 1.97 and $2.31\mu_B$ [305]. Furthermore, the magnetic transition temperature in alloys with Ni level less than $\sim$ 30% was found to be below 150 K [294]. The above experimental evidence implies that at room temperature the FeCrNi alloys are paramagnetic with sizable persisting disordered local magnetic moments. This is in contrast to the basic stainless steels made of Fe and Cr, which are ferromagnetic at ambient condition with a Curie temperature of ~900–1050 K [262].

In the theoretical description of alloy steels using the EMTO method, the paramagnetic ternary $Fe_{1-c-n}Cr_cNi_n$ system is modeled by an *fcc* alloy with randomly distributed chemical species and local magnetic moments, *i.e.* by a quasi-ternary random

$$\mathrm{Fe}^{\uparrow}_{(1-c-n)/2}\mathrm{Fe}^{\downarrow}_{(1-c-n)/2}\mathrm{Cr}^{\uparrow}_{c/2}\mathrm{Cr}^{\downarrow}_{c/2}\mathrm{Ni}^{\uparrow}_{n/2}\mathrm{Ni}^{\downarrow}_{n/2}$$

alloy. Here the arrows represent the two magnetic moments oriented up ($\uparrow$) and down ($\downarrow$). Often this approximation is referred to as the *disordered local moment* (DLM) approach, which accurately describes the effect of loss of the net magnetic moment above the transition temperature [260, 261]. In fact, at $T = 0$ K, the local magnetic moments on Cr and Ni sites vanish, and therefore the above expression reduces to

$$\mathrm{Fe}^{\uparrow}_{(1-c-n)/2}\mathrm{Fe}^{\downarrow}_{(1-c-n)/2}\mathrm{Cr}_c\mathrm{Ni}_n$$

However, at finite temperature, the magnetic entropy induces magnetic fluctuations on Cr and Ni sites, which have to be accounted for by an accurate description of the energetics of FeCrNi random alloys [263]. When a fourth alloying element M is introduced, the quasi-ternary alloy becomes a quasi-quaternary random alloy, *viz.*

$$\mathrm{Fe}^{\uparrow}_{(1-c-n-m)/2}\mathrm{Fe}^{\downarrow}_{(1-c-n-m)/2})\mathrm{Cr}^{\uparrow}_{c/a}\mathrm{Cr}^{\downarrow}_{c/a}\mathrm{Ni}^{\uparrow}_{n/2}\mathrm{Ni}^{\downarrow}_{n/2}\mathrm{M}^{\uparrow}_{m/2}\mathrm{M}^{\downarrow}_{m/2}$$

where m stands for the atomic fraction of M. Hence, for an adequate theoretical simulation of alloy steels both the chemical and magnetic disorder should be treated simultaneously. This is within the reach of the EMTO method. The special feature of this approach to correctly account for anisotropic lattice distortions in multi-component random systems offers unique possibilities for describing the alloy steels at quantum mechanical level.

9.2 Elastic Properties of Alloy Steels

9.2.1 Elastic Constants of FeCrNi Alloys

In Table 9.2, the calculated single-crystal elastic constants, the Zener elastic anisotropy ratio and the poly-crystalline averaged engineering elastic mod-

Table 9.2. Theoretical (EMTO-GGA) and experimental equilibrium volume (V), single-crystal elastic constants (c_{11}, c_{12}, and c_{44}), Zener elastic anisotropy constant (A), polycrystal bulk modulus (B), shear modulus (G) and B/G ratio for $Fe_{0.72}Cr_{0.20}Ni_{0.08}$, corresponding approximately to alloy steel AISI 304. The volume is in Bohr^3 and the elastic moduli in GPa. The experimental data are from [306] for the volume and [55] for the elastic constants. In the third row we give the relative deviations (Δ in percent) between the EMTO and experimental data.

	V	c_{11}	c_{12}	c_{44}	A	B	G	B/G
EMTO	78.93	208.6	143.5	132.8	4.07	165.2	77.3	2.13
expt.	79.45	204.6	137.7	126.2	3.78	158.2	77.4	2.04
Δ	-0.6	2.0	4.2	5.2	7.6	4.4	-0.1	4.4

uli are compared with the available experimental data for $Fe_{0.72}Cr_{0.20}Ni_{0.08}$, corresponding approximately to stainless steel AISI 304 (see Table 9.1). It is found that the EMTO-GGA approach[1] reproduces the experimental equilibrium volume [306] within less than 1%. Note that this error is close to those obtained for elemental Fe, Cr and Ni in conjunction with the GGA functional. The average deviation between the EMTO and experimental elastic constants is about 4%, which is below the typical errors obtained for ele-

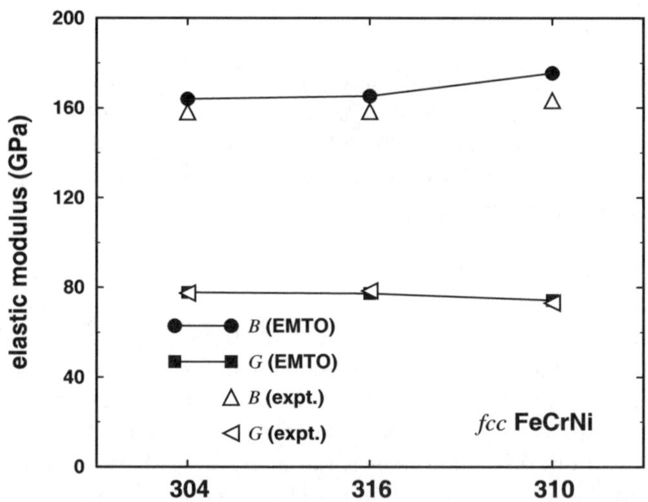

Fig. 9.1. Comparison between EMTO and experimental [307, 308, 309] shear and bulk moduli of alloy steels AISI 304: $Fe_{0.714}Cr_{0.197}Ni_{0.089}$, 316: $Fe_{0.692}Cr_{0.186}Ni_{0.122}$ and 310: $Fe_{0.544}Cr_{0.262}Ni_{0.194}$. The experimental values correspond to commercial stainless steels from Table 9.1 but without Mo for steel 316.

[1] Since for the 3d metals the GGA functionals outperforms the other existing density functional approximations (see Table 7.1), all the theoretical calculations for steels were performed within the GGA.

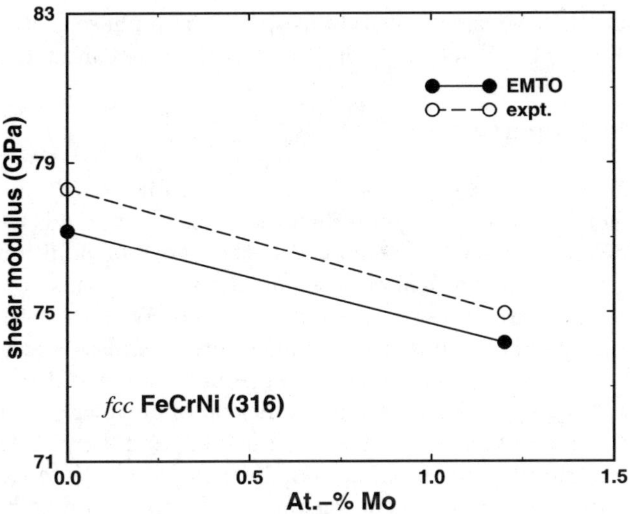

Fig. 9.2. The effect of Mo addition on the shear modulus of austenitic stainless steel 316: $Fe_{0.692-m}Cr_{0.186}Ni_{0.122}Mo_m$. The EMTO results are compared with the experimental data [307].

mental 3*d* transition metals [188]. It is also significant that the extremely high elastic anisotropy of austenite is very well captured by the theory. The isotropic shear modulus from Table 9.2 was derived from the single-crystal data using the Hershey averaging method (Section 6.3.1). We find that the theoretical results for both B and G as well as for the ratio between them are in excellent agreement with experiment. Therefore, one concludes that the EMTO-GGA scheme can reproduce with high accuracy the individual bulk parameters of the FeCrNi system.

Since there is no restriction regarding the number of alloy components and concentrations, the EMTO method allows for the determination of the effect of alloying on the elastic properties. The theoretical polycrystalline elastic moduli for some selected FeCrNi alloys, for which experimental data are available [307, 308, 309], are shown in Figure 9.1. The average deviation between the theoretical and experimental shear and bulk moduli from figure are 1.0% and 5.1%, respectively. In Figure 9.2, the effect of alloying with Mo on the shear modulus of grade 316 is shown. Both theory and experiment predict a substantial decrease in G.

It is interesting to compare the effect of Mo addition on the shear modulus with the composition dependence of the stability of the austenitic (*fcc*) phase relative to the ferrite phase, which has the *bcc* crystallographic structure. We recall that in the case of transition metals the cubic elastic constant associated with the tetragonal distortion of the lattice, shows a proportionality to the energy difference between the *bcc* and *fcc* structures [188]. In the case of alloy

316, it is found that Mo strongly stabilizes the ferrite phase (~ 0.22 mRy per At.-%), which correlates well with the trend of the shear modulus.

9.2.2 Elastic Property Maps

The most significant conclusion from Table 9.2 and Figures 9.1 and 9.2 is that the EMTO method accurately reproduces the observed trends of the elastic moduli of FeCrNi ternary alloys and also the effect of additional alloying elements. Therefore, theoretical results calculated using this computational tool may be used for prediction of new data on steels. With this object, a series of composition–elastic moduli maps of austenitic stainless steels have been created [105]. These maps were constructed from the bulk and shear modulus of $Fe_{1-c-n}Cr_cNi_n$ alloys calculated as a function of chemical composition for $0.135 < c < 0.255$ and $0.08 < n < 0.24$. Figures 9.3 and 9.4 give the chemical composition distribution of the bulk and shear modulus. Note the difference between the orientations of the concentration (Cr and Ni) axis on these two figures. The maps were generated for a concentration interval that includes the basic compositions of the well-known commercial stainless steels (Table 9.1).

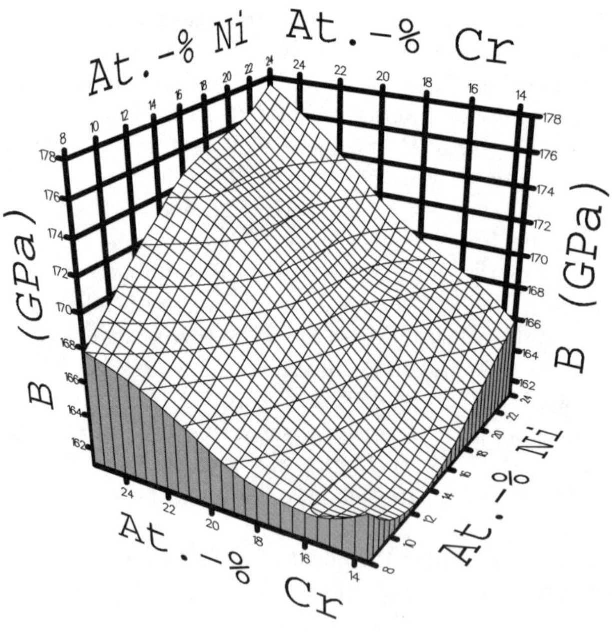

Fig. 9.3. Calculated (EMTO-GGA) bulk modulus of FeCrNi alloys as a function of Cr and Ni contents (balance iron).

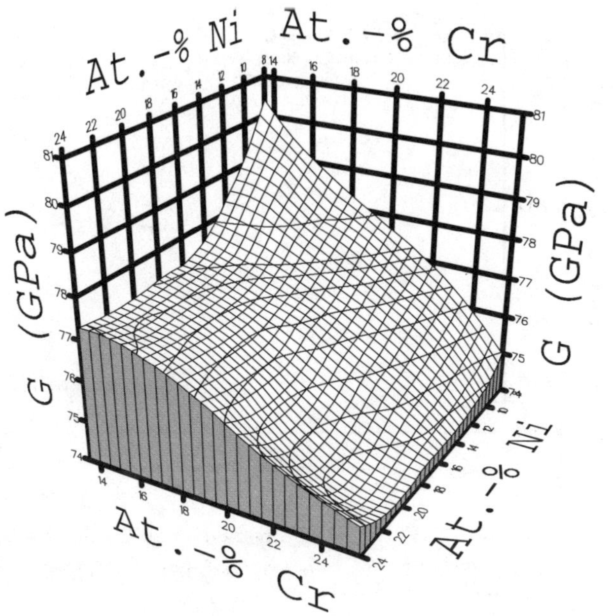

Fig. 9.4. Calculated (EMTO-GGA) shear modulus of FeCrNi alloys as a function of Cr and Ni contents (balance iron).

The bulk modulus of FeCrNi alloys (Figure 9.3) varies between a minimum value of 161 GPa, corresponding to $Fe_{0.75}Cr_{0.13}Ni_{0.12}$, and a maximum value of 178 GPa, belonging to $Fe_{0.51}Cr_{0.25}Ni_{0.24}$. It follows from figure that both Cr and Ni enhance the bulk modulus of alloy steel. In terms of the shear modulus (Figure 9.4) three families of alloys can be distinguished. Compounds with large shear modulus correspond to low and intermediate Cr ($< 20\%$) and low Ni ($< 15\%$) concentrations. Within this group of alloys G decreases monotonically with both Cr and Ni from a pronounced maximum about 81 GPa (near $Fe_{0.78}Cr_{0.14}Ni_{0.08}$) to approximately 77 GPa. The high Cr content alloys define the second family of austenites possessing the lowest shear moduli (≤ 75 GPa) with a minimum around the composition $Fe_{0.755}Cr_{0.25}Ni_{0.20}$. The third family of austenites, with intermediate G values, is located at moderate Cr ($< 20\%$) and high Ni ($> 15\%$) concentrations, where G shows no significant chemical composition dependence.

According to the equation of state (Section 6.1), lower atomic volume corresponds to larger bulk modulus and vice versa. Tracing the alloying effects on the theoretical equilibrium volume (Figure 9.5) we find that both Cr and Ni addition increase the average lattice constant of paramagnetic FeCrNi alloys. This trend is just the opposite of the trend obeyed by the bulk modulus

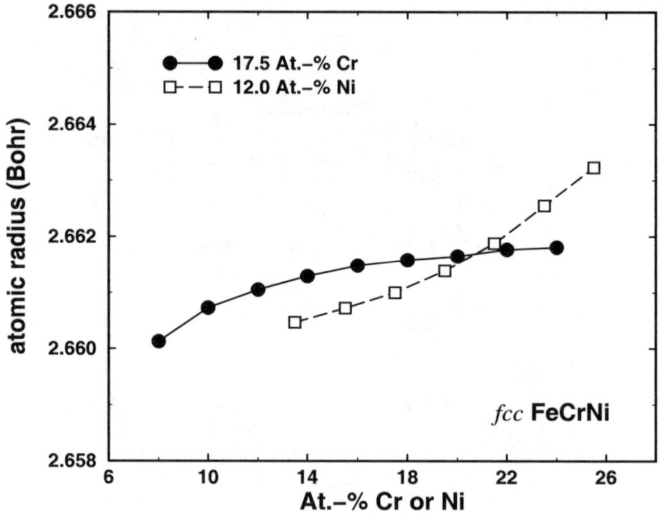

Fig. 9.5. Calculated (EMTO-GGA) average equilibrium atomic radius of $Fe_{1-0.175-n}Cr_{0.175}Ni_n$ (circles) and $Fe_{1-c-0.12}Cr_cNi_{0.12}$ (squares) alloys as a function of Ni and Cr content, respectively.

from Figure 9.3, which indicates that the alloying effect on the bulk modulus of FeCrNi alloys is determined first of all by the changes in the electronic structure.

The effect of Cr and Ni on G may be contrasted with the composition dependence of the stability of the austenite (fcc) phase relative to the ferrite phase. EMTO total energy calculations carried out for the fcc and bcc phases show that Ni always stabilizes the austenitic phase (with $\sim$ 0.07 mRy per At.-%). However, even so, except for large concentrations, it decreases the cubic and thus the polycrystalline elastic moduli. On the other hand, Cr is a strong ferrite stabilizer ($\sim$ 1.0 mRy per At.-%), which correlates reasonably well with the trend of the shear modulus from Figure 9.4.

9.2.3 Quaternary FeCrNi-based Alloys

The effect of quaternary alloying additions has been investigated in the case of $Fe_{0.58}Cr_{0.18}Ni_{0.24}$ [246, 310]. The EMTO calculations were performed for $Fe_{0.58-x}Cr_{0.18}Ni_{0.24}M_x$ alloys with x =0–0.05 At.-%. Here M denotes Al, Si, V, Cu, Nb, Mo, Re, Os, or Ir. All of these elements are known to be useful alloying agents in commercial steels.

In Figures 9.6 and 9.7 we show the EMTO results for the average effects (per atomic percent) of alloying additions on the equilibrium volume, phase stability and polycrystalline elastic moduli of $Fe_{0.58-x}Cr_{0.18}Ni_{0.24}M_x$ alloys. We find that all metallic alloying additions increase, while Si slightly decreases

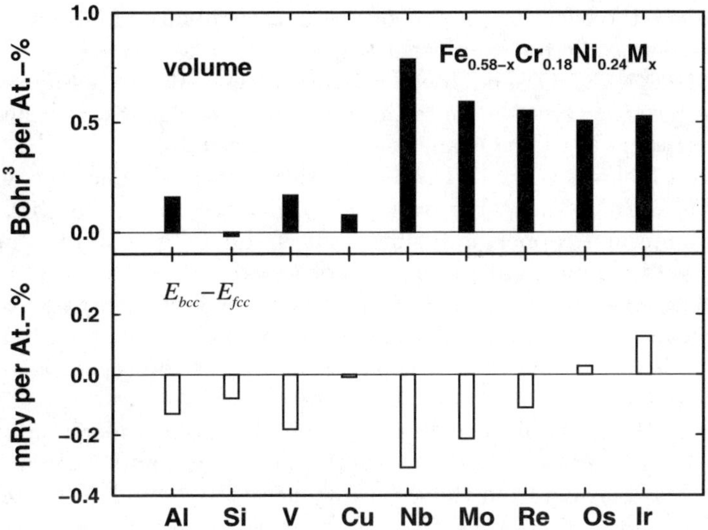

Fig. 9.6. The effect of alloying additions (shown on the abscissa) on the equilibrium volume (upper panel, in Bohr^3) and structural energy difference between *bcc* and *fcc* phases (lower panel, in mRy) for $Fe_{0.58-x}Cr_{0.18}Ni_{0.24}M_x$ random alloys.

the equilibrium volume of high-Ni austenite. The effect of Nb and Mo compares well with the experimental observations [311, 312]. Among the 5*d* metals Os has the smallest and Re the largest effect, which correlates with the equilibrium volumes of pure Re, Os and Ir. However, the average increase in the lattice parameter in the case of 5*d* elements is ~ 0.011 Bohr per At.-%, which is significantly larger than the one estimated from the Vegard's rule.

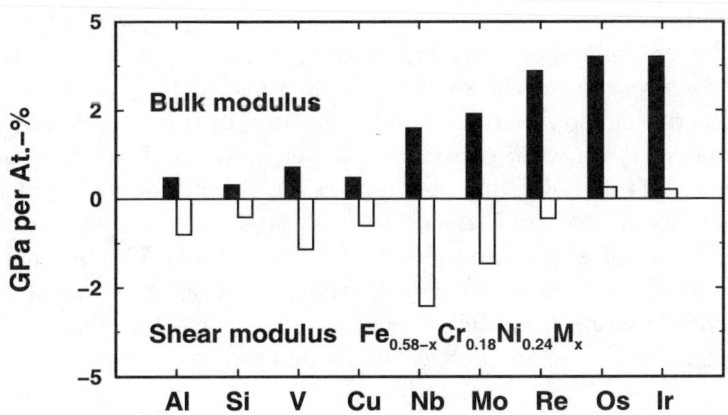

Fig. 9.7. The effect of alloying additions (shown on the abscissa) on the bulk and shear moduli (in GPa) for $Fe_{0.58-x}Cr_{0.18}Ni_{0.24}M_x$ random alloys.

We assess the effect of alloying on the phase stability of $Fe_{0.58}Cr_{0.18}Ni_{0.24}$ alloy by comparing the total energies calculated for ferrite and austenite phases. Except Cu, Os and Ir, all the other elements from Figure 9.6 stabilize the *bcc* structure relative to the *fcc* structure. The energy difference for Cu is close to zero, while Os and Ir are austenite stabilizers like Ni.

Figure 9.7 shows that all elements considered here enhance the bulk modulus of $Fe_{0.58}Cr_{0.18}Ni_{0.24}$. For example, Si increases B by 0.6 GPa per At.-%. Aluminum and copper have similar effects on B. A somewhat more pronounced bulk modulus enhancement is observed in the case of $4d$ transition metals. The effect of $5d$ metals is almost twice as large compared to that of the $4d$ metals. The bulk modulus enhancing effect of the *bcc* transition metals (V, Nb and Mo) in the case of commercial steel grades (*e.g.*, in AISI 316) has been pointed out also in experiments [307]. The physical origin of the variation of B with additional transition metals can be understood using the Friedel model of cohesion in d metals [313] and taking into account the bond enhancement as going from the third-row metals to $4d$ and $5d$ metals [246].

The shear modulus of $Fe_{0.58}Cr_{0.18}Ni_{0.24}$ shows no significant variation with Os and Ir contents, and it is slightly diminished with Si, Cu and Re additions. At the same time, Al, V, Nb and Mo decrease G significantly. Comparing the lower panels from Figures 9.7 and 9.6, one can see that the variation of G correlates well with the variation of the structural energy difference between the *bcc* and *fcc* phases. In particular, both Os and Ir stabilize the austenite phase and also increase the shear modulus of alloy steel. The fact that the increase of G is relatively small is due to the simultaneous increase of the average atomic volume upon alloying. The negative volume derivative of the shear modulus diminishes the increase in G upon alloying with Os or Ir.

The chemical composition dependence of G can be understood by investigating the nature of the crystal bonds. The shear modulus represents the opposition of a bulk material against shear, *i.e.* bond rotation. Hence, strongly anisotropic bonds correspond to larger shear modulus, while isotropic crystals have lower shear modulus. The nature of bonding can be characterized, *e.g.*, by the anisotropy of the surface free energies [171]. In systems where the valence electron density is distributed uniformly in the interstitial region, the surface energy shows week orientation dependence. This kind of behavior is typical for the Ti and V group elements [171]. Directional or covalent type of bonding, on the other hand, gives rise to strongly anisotropic surface energy, which is the case of middle and late transition metals [171]. In particular, Re, Os and Ir have been found to present the largest surface energy anisotropy amongst the d metals. The directional bonding characters in Mo and Cu are similar to that observed in the case of Cr and Ni, respectively.

9.2.4 Misfit Parameters of FeCrNiM Alloys

The size and elastic misfit parameters for solute M in alloy with composition $Fe_{0.58}Cr_{0.18}Ni_{0.24}$ can be determined using the EMTO data plotted in Figures

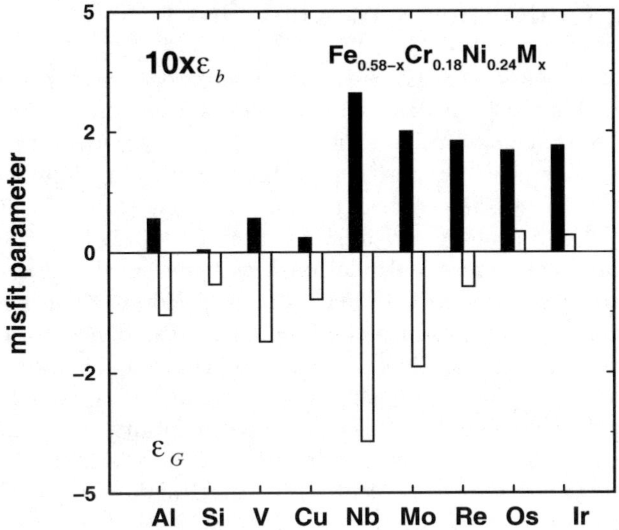

Fig. 9.8. Theoretical (EMTO-GGA) misfit parameters for $Fe_{0.58}Cr_{0.18}Ni_{0.24}$ alloy comprising a few percent of Al, Si, V, Cu, Nb, Mo, Re, Os or Ir. Note that the size misfit parameters have been multiplied by a factor of 10.

9.6 (upper panel) and 9.7 (lower panel). Figure 9.8 gives ε_b and ε_G for Al, Si, V, Cu, Nb, Mo, Re, Os and Ir. The two misfit parameters are calculated according to Equation (6.111) and they are obtained by dividing the data from Figures 9.6 and 9.7 by the equilibrium volume and shear modulus of $Fe_{0.58}Cr_{0.18}Ni_{0.24}$, respectively. The size misfit is negligible for Al, Si, V and Cu, but it has a sizable value (between 0.21 and 0.33) for the $4d$ and $5d$ dopants. It is found that Al, V, Nb and Mo give elastic misfit parameters $|\varepsilon_G| = 1.3$, 1.9, 3.9 and 2.4, respectively, whereas $|\varepsilon_G| < 1$ for Si, Cu, Re, Os and Ir.

The Fleischer parameter ε_L is 5.4 for Nb, 4.1 for Mo, and ~ 3.5 for the $5d$ elements. All the other dopants give $\varepsilon_L < 1.5$. Hence, assuming that the Labusch–Nabarro model (see Section 6.6) is valid in the case of FeCrNi alloys encompassing a few percent of substitutional elements, Nb and Mo are predicted to yield the largest solid solution hardening. However, one should also take into account that the $4d$ metals, in contrast to the $5d$ metals, significantly decrease G. Since the Peierls stress is approximately proportional to the shear modulus [148], the $\sim 2-3$ GPa per At.-% decrease in G (Figure 9.7) in the case of Nb and Mo is expected to diminish the corresponding Peierls stress as well. Therefore, the overall hardening effect, obtained as a sum of the solid solution hardening and the Peierls term, might be somewhat different from the one expressed merely via the Fleischer parameter.

9.3 Stacking Fault Energy of Alloy Steels

The stacking fault energy (SFE) is a key microscopic parameter of the austenitic phase. The fact that the SFE can be subtly altered by alloying elements indicates that the mechanical properties of stainless steels can be controlled by the chemical composition [314]. The stacking fault energy in austenitic steels has been determined from experiment [155, 154, 315, 316, 317, 318, 319, 320] and semi empirical simulation [318, 321, 322, 323]. Using the so derived databases, empirical relationships between SFE and chemical compositions have been established [154, 315, 320]. However, in most cases, these relationships fail to reproduce the nonlinear dependence obtained, for instance, in high-Ni alloys [315, 263], or the alloying effects in hosts with different compositions. The failure of such parameterizations is partially due to the fact that the SFE is difficult to measure experimentally and usually large inaccuracies are associated with the values quoted from the literature.

Using the EMTO method, we presented a theoretical database of the stacking fault energy (γ) of FeCrNi random alloys [263, 324]. These calculations were carried out for the $Fe_{1-c-n}Cr_cNi_n$ alloys, with $0.135 < c < 0.255$ and $0.08 < n < 0.20$. The SFE was computed according to Equation (6.110). Because of the large magnetic fluctuations in these alloys [263], the local magnetic moments have to be taken into account for all alloy components. To this end, for each individual atom from the unit cell we introduced a local magnetic moment μ_i, where i stands for Fe, Cr and Ni. These moments were determined as a function of temperature (T) from the minimum of the free energy $F(T,\mu) = E(T,\mu) - T[S_{\mathrm{mag}}(\mu) + S_{\mathrm{el}}(T)]$, calculated as a function of temperature and local magnetic moments. Here μ denotes the set of local atomic moments μ_{Fe}, μ_{Cr}, *etc.* In actual calculations, the electronic energy $E(T,\mu)$ and electronic entropy $S_{\mathrm{el}}(T) = -2k_{\mathrm{B}} \int \{f(\epsilon)\ln f(\epsilon) + [1-f(\epsilon)]\ln[1-f(\epsilon)]\} D(\epsilon) d\epsilon$ (where $D(\epsilon)$ is the density of states) were obtained from spin-constrained EMTO calculations, using the finite-temperature Fermi distribution $f(\epsilon)$. In fixed-spin calculations, we applied a constant splitting for all the atoms from the unit cell. The theoretical equilibrium volume was calculated from the 0 K electronic energy. The magnetic entropy $S_{\mathrm{mag}}(\mu)$ for atom i was estimated using the mean-field expression $k_{\mathrm{B}} log(\mu_i + 1)$ valid for completely disordered localized moments [297].

9.3.1 Theoretical Stacking Fault Energy versus Experimental Data

In Figures 9.9 and 9.10, we compare the room-temperature theoretical stacking fault energies for $Fe_{1-c-n}Cr_cNi_n$ alloys with the available experimental data [155, 154, 318, 317, 319]. In figures, the SFE is plotted as a function of n for alloys containing 17 and 19% Cr (Figures 9.9), and as a function of c for alloys containing 14 and 16% Ni (Figures 9.10). The experimental values were obtained for alloys containing small amounts of interstitial (*e.g.*, C, N) and other substitutional (*e.g.*, Mo, Si, Mn) elements as well. Taking into account

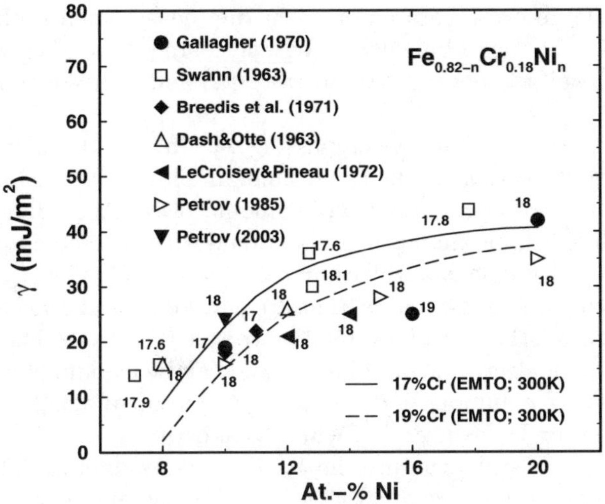

Fig. 9.9. Theoretical (EMTO-GGA) and experimental SFE energy of FeCrNi alloys as a function of Ni content. References for experiments [155, 317, 154, 318, 319] are given in legend. The numbers next to the symbols give the actual compositions used in experiments.

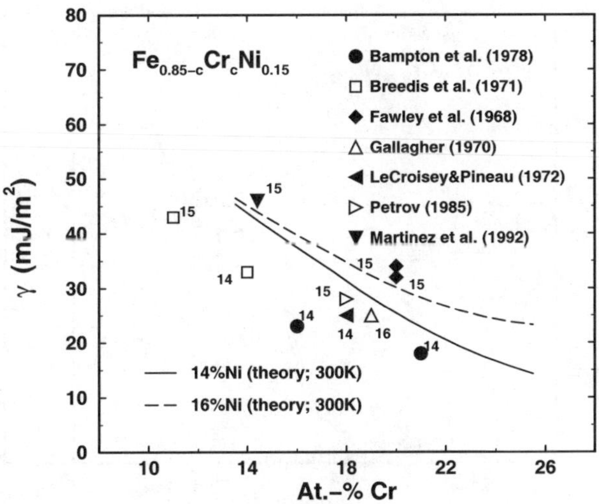

Fig. 9.10. Theoretical (EMTO-GGA) and experimental SFE energy of FeCrNi alloys as a function of Cr content. References for experiments [155, 317, 154, 318, 319] are given in legend. The numbers next to the symbols give the actual compositions used in experiments.

the relatively large error bars reported in the measurements, the agreement between theory and experiment can be considered good. We would like to point out how well the observed trends for $\gamma(n)$ and $\gamma(c)$ are captured by the present theory.

It is found that, in alloys encompassing 14–16% Ni, Cr always decreases the stacking fault energy. At the same time, in alloys with $c = 0.17-0.19$, the stacking fault energy increases with the amount of Ni. According to Equation (6.109), the SFE is approximately proportional with the energy difference between the *hcp* and *fcc* phases. Hence, in terms of structural energy differences, the trends from Figures 9.9 and 9.10 mean that in FeCrNi alloys Cr stabilizes and Ni destabilizes the *hcp* phase relative to the *fcc* phase. These trends are partially due to an electronic mechanism, and can be explained using the variation of the effective number of d electrons (N_d) upon alloying. We recall that Cr substitution for Fe reduces N_d, while Ni substitution for Fe increases N_d. According to the crystal structure theory of transition metals [325], the structural energy difference approximately scales with the difference in the density of states $D(\epsilon)$ near the Fermi level ϵ_F. EMTO calculations show that in *fcc* alloys $D^{fcc}(\epsilon)$ is almost constant near ϵ_F (Figure 9.11), and, thus, alloying with Cr or Ni produces no significant effect on $D^{fcc}(\epsilon_F)$. At the same time, in *hcp* alloys $D^{hcp}(\epsilon)$ presents a pronounced minima at energies slightly below ϵ_F. Therefore, a small amount of Ni (Cr) addition is expected to destabilize (stabilize) the *hcp* structure. This explains the trends from Figures 9.9 and 9.10.

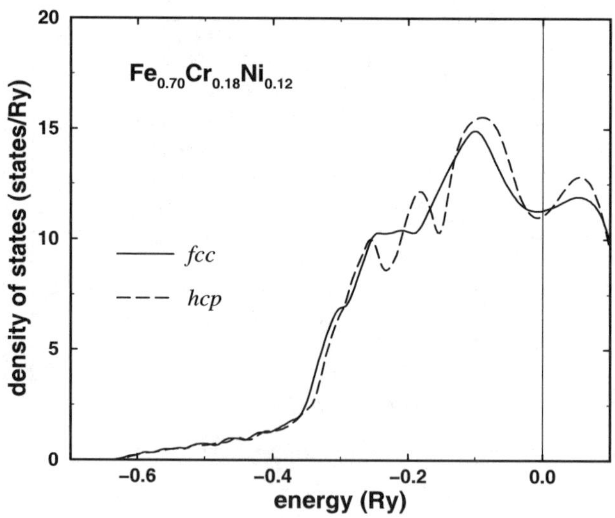

Fig. 9.11. Density of states for $Fe_{0.8}Cr_{0.18}Ni_{0.12}$ random alloy calculated for *fcc* (solid line) and *hcp* (dashed line) phases.

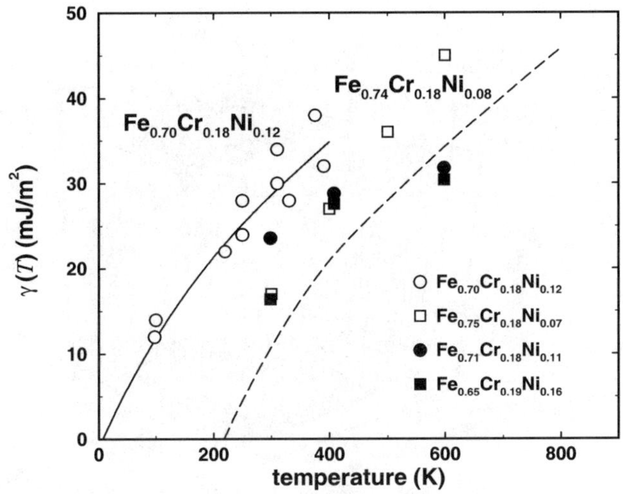

Fig. 9.12. Temperature dependence of the SFE. Lines denote the EMTO results for $Fe_{0.70}Cr_{0.18}Ni_{0.12}$ (solid line) and $Fe_{0.74}Cr_{0.18}Ni_{0.08}$ (dashed line). Symbols are the experimental values for $Fe_{0.70}Cr_{0.18}Ni_{0.12}$ [327], $Fe_{0.71}Cr_{0.18}Ni_{0.11}$ and $Fe_{0.65}Cr_{0.19}Ni_{0.16}$ [326], and the average experimental values for $Fe_{0.75}Cr_{0.18}Ni_{0.07}$ [328] with errors of 20 mJ/m^2.

The calculated temperature dependence of the SFE for two compositions, close to those considered in the experiments [326, 327, 328], is shown in Figure 9.12. The large (about 10 mJ/m^2 per 100 K) SFE versus temperature slope obtained in the experiments is very well reproduced by the EMTO method. For all data, $\partial\gamma/\partial T$ slightly decreases with T, showing a tendency to saturate at high temperatures.

9.3.2 Magnetic Stacking Fault Energy

The electronic mechanism for the alloying effects on the SFE, described in Section 9.3.1, is restricted to (i) compositions, where the magnetic contribution to the SFE is small, and (ii) alloying elements, which leave unchanged the equilibrium volume of the host. However, in a process where the magnetic structure and/or the volume is also modified, there are other mechanisms which play important roles in the composition induced changes in the SFE. This is in fact the situation in low-Cr and high-Ni alloys. In Figure 9.13, we show the calculated room-temperature SFE of FeCrNi alloys as a function of the chemical composition. The observed strongly nonlinear composition dependence is a consequence of the persisting local magnetic moments in austenitic steels.

In order to understand this behavior we divide the stacking fault energy into a 0 K contribution, γ_0, plus the remaining part, $\gamma_{\rm mag}$, *viz.* $\gamma = \gamma_0 + \gamma_{\rm mag}$. Because at room temperature the Fermi function from the electronic energy

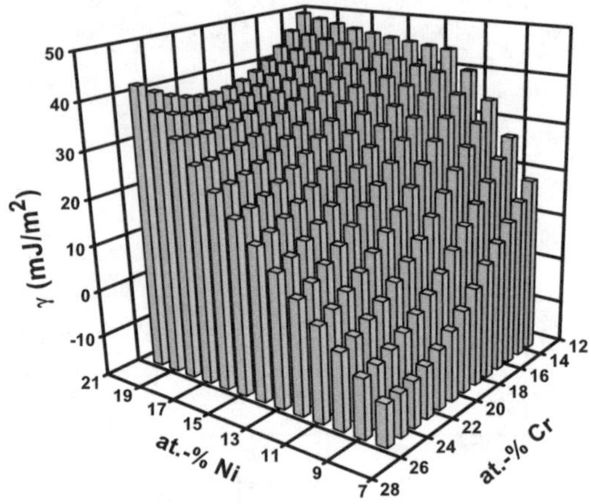

Fig. 9.13. Stacking fault energy of FeCrNi random alloys calculated using the EMTO-GGA scheme. The SFE is shown for T = 300 K as a function of Cr and Ni content.

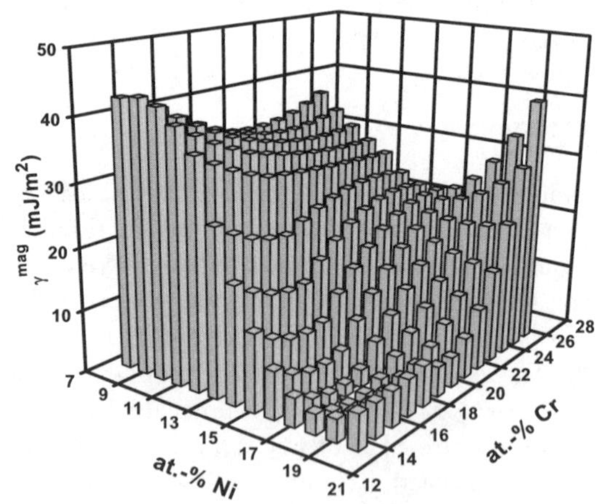

Fig. 9.14. Magnetic fluctuation contribution to the stacking fault energy from Figure 9.13. γ^{mag} is plotted for $T = 300$ K as a function of chemical composition.

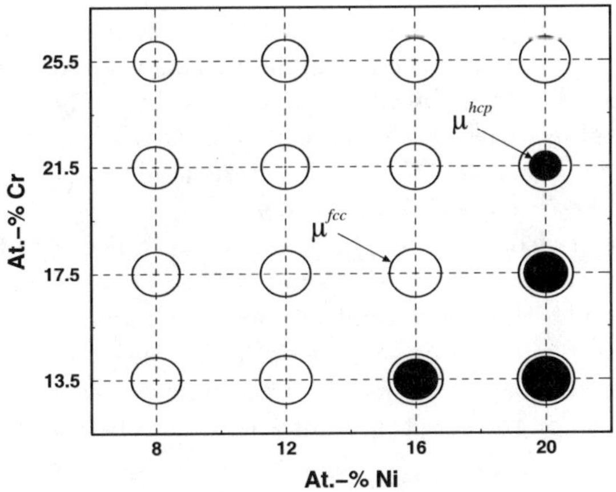

Fig. 9.15. Theoretical (EMTO) local magnetic moments in *fcc* (open circles) and *hcp* (filled circles) FeCrNi alloys calculated at $T = 0$ K as a function of chemical composition. The size of the spheres is proportional with the magnitude of the magnetic moments. The large spheres at 13.5% Cr and 20% Ni correspond to $\mu^{fcc} = 1.75\mu_B$ and $\mu^{hcp} = 1.48\mu_B$, and the small spheres at 25.5% Cr and 8% Ni represent $\mu^{fcc} = 1.33\mu^B$ and $\mu^{hcp} = 0.00\mu_B$.

and entropy is relatively insignificant, the dominant temperature dependence of the free energy comes from the $-\mathrm{T}S_{\mathrm{mag}}$ term. Hence, we identify γ_0 with the chemical, and γ_{mag} with the magnetic fluctuation [263] contribution to the SFE. The local magnetic moments in the double hexagonal structure are calculated to be close to those in the *fcc* structure. Therefore, according to Equation (6.109), the dominant part of γ_{mag} can be expressed as $-\mathrm{T}(S_{\mathrm{mag}}^{hcp} - S_{\mathrm{mag}}^{fcc})/A_{2\mathrm{D}}$. In other words, in alloys where the local magnetic moments from the *hcp* (μ^{hcp}) and *fcc* (μ^{fcc}) phases differ significantly one can expect a large magnetic contribution to the SFE.

Theoretical room-temperature data for γ_{mag} are shown in Figure 9.14. We find that the magnetic SFE has the same order of magnitude as the total SFE from Figure 9.13, confirming the importance of the disordered local moments for the stability of steels. For the entire composition interval considered here, the paramagnetic *fcc* alloys have large disordered local moments. At 0 K these moments are located on Fe atoms, and have magnitudes per Fe atom ranging from $1.35\mu_B$, near $Fe_{67}Cr_{25}Ni_8$, to $1.75\mu_B$, near $Fe_{67}Cr_{13}Ni_{20}$. At theoretical volumes, the low-Ni *hcp* alloys are calculated to be non-magnetic with vanishing local magnetic moments. This is illustrated in Figure 9.15. However, the low γ_{mag} values found for the low-Cr and high-Ni corner of Figure 9.14, indicate that the alloys from this corner of the map have similar S_{mag} in the *hcp* and *fcc* phases. Indeed, we find that upon Ni addition the low-Cr alloys undergo a transition from a non-magnetic to a paramagnetic phase (Figure 9.15) with non-vanishing local moments. At high-Ni and low-

Cr contents the *hcp* alloys possess disordered magnetic moments of 0.95–1.50 μ_B per Fe atom. In contrast to Ni, calculations show that Cr addition always tends to stabilize the zero local moment solution in the hexagonal phase. For instance, in *hcp* alloys with $n \approx 0.18-0.20$, the local moments disappear with c increasing from 0.135 to 0.255. This magnetic transition explains the change obtained in the $\partial\gamma/\partial c$ slope from Figure 9.13.

The presence of disordered magnetic moments in FeCrNi alloys has important implications for the alloying effects on the SFE of quaternary alloys. It has been shown [324] that as a result of magnetic transition in the *hcp* phase, the same alloying element can cause totally opposite changes in the SFE of alloys with different host composition. This indicates that no universal composition equations for the stacking fault energy can be established.

Part III

Auxiliary Material

Appendix A

Exchange-correlation Approximations

In this appendix, we shall present a few expressions for the exchange-correlation functionals used in the Density Functional calculations from this book. In the case of the Local Density approximation (LDA), there is a series of different parametrizations [4, 5, 7] of the Monte Carlo data on the electron–electron interaction by Ceperley and Alder [6]. Here we shall give the one established by Perdew and Wang [7], which is analytic as a function of the density parameter and relative spin polarization. For the gradient level approximations, two functionals will be considered: the Generalized Gradient approximation (GGA) by Perdew and co-workers [9, 16, 15], and the Local Airy Gas approximation (LAG) by Vitos and co-workers [17].

The exchange-correlation energy is written in the form

$$E_{xc}[n] = E_x[n] + E_c[n] = \int n(\mathbf{r})\varepsilon_x([n];\mathbf{r})d\mathbf{r} + \int n(\mathbf{r})\varepsilon_c([n];\mathbf{r})d\mathbf{r}, \quad \text{(A.1)}$$

where $\varepsilon_x([n];\mathbf{r})$ and $\varepsilon_c([n];\mathbf{r})$ represent the exchange and correlation energies per electron, respectively. In spin polarized case, with $n_\downarrow$ and $n_\uparrow$ spin-down and spin-up electron densities, the above functionals depend on both spin densities. In particular, the exchange energy obeys the following scaling relation [329]

$$E_x[n_\uparrow, n_\downarrow] = (E_x[2n_\uparrow] + E_x[2n_\downarrow])/2. \quad \text{(A.2)}$$

Self-consistent Density Functional calculations require the exchange-correlation part of the Kohn-Sham potential (1.4). This is obtained as the functional derivative of the exchange-correlation energy (A.1). Using the Full Charge Density technique, described in Chapter 4, the self-consistent calculation is always carried out within the Local Density approximation, and the output total density is used to calculate the total energy at different gradient level approximations. Therefore, here $\mu_{xc}([n];\mathbf{r}) \equiv \delta E_{xc}[n]/\delta n(\mathbf{r})$ is given only within the LDA. The GGA and LAG level exchange potentials can be computed using the equations from Perdew and Wang [12], while the GGA level correlation potential can be found in, *e.g.*, Perdew *et al.* [330].

A.1 Local Density Approximation

For a uniform non-polarized electron gas, the exchange energy per electron is

$$\varepsilon_x^{\mathrm{LDA}}(n) = -\frac{3}{2}\left(\frac{3}{\pi}\right)^{1/3} n^{1/3}. \tag{A.3}$$

The exchange potential is the density derivative of $n\varepsilon_x^{\mathrm{LDA}}(n)$, *i.e.*

$$\mu_x^{\mathrm{LDA}}(n) = -2\left(\frac{3}{\pi}\right)^{1/3} n^{1/3}. \tag{A.4}$$

The spin polarized exchange energy $\varepsilon_x^{\mathrm{LDA}}(n_\uparrow, n_\downarrow)$ and the corresponding potentials $\mu_{x,\sigma}^{\mathrm{LDA}}(n_\uparrow, n_\downarrow) = \partial\varepsilon_x^{\mathrm{LDA}}(n_\uparrow, n_\downarrow)/\partial n_\sigma$ ($\sigma =\uparrow$ or $\downarrow$) are obtained using the scaling relation (A.2).

For a uniform polarized electron gas, the correlation energy is expressed in terms of the density parameter r_s [7],

$$r_s = \left[\frac{3}{4\pi(n_\uparrow + n_\downarrow)}\right]^{1/3}, \tag{A.5}$$

and the relative spin polarization η,

$$\eta = \frac{n_\uparrow - n_\downarrow}{n_\uparrow + n_\downarrow}, \tag{A.6}$$

and it has the following form

$$\begin{aligned} \varepsilon_c^{\mathrm{LDA}}(r_s, \eta) &= \varepsilon_c(r_s, 0) + \alpha_c(r_s)\frac{f(\eta)}{f''(0)}(1-\eta^4) + \\ &+ \left[\varepsilon_c(r_s, 1) - \varepsilon_c(r_s, 0)\right] f(\eta)\eta^4, \end{aligned} \tag{A.7}$$

where

$$f(\eta) = \frac{(1+\eta)^{4/3} + (1-\eta)^{4/3} - 2}{2^{4/3} - 2}. \tag{A.8}$$

In (A.7), the functions $\varepsilon_c(r_s, 0), \varepsilon_c(r_s, 1)$ and $\alpha_c(r_s)$ are given by

$$\begin{aligned} G(r_s, A, \alpha_1, \beta_1, \beta_2, \beta_3, \beta_4) &= -4A(1+\alpha_1 r_s) \times \\ \times \ln\left[1 + \frac{1}{2A(\beta_1 r_s^{1/2} + \beta_2 r_s + \beta_3 r_s^{3/2} + \beta_4 r_s^2)}\right], \end{aligned} \tag{A.9}$$

with $\alpha_1, A, \beta_{1,\ldots,4}$ listed in Table A.1. The exchange potential for spin σ is

Table A.1. Parameters for the correlation energy [7].

	$\varepsilon_c(r_s,0)$	$\varepsilon_c(r_s,1)$	$-\alpha_c(r_s)$
A	0.031091	0.015545	0.016887
α_1	0.21370	0.20548	0.11125
β_1	7.5957	14.1189	10.357
β_2	3.5876	6.1977	3.6231
β_3	1.6382	3.3662	0.88026
β_4	0.49294	0.62517	0.49671

$$\begin{aligned}\mu_{c,\sigma}^{\mathrm{LDA}}(r_s,\eta) &= \frac{\partial \varepsilon_c^{\mathrm{LDA}}(r_s,\eta)}{\partial n_\sigma} = \varepsilon_c^{\mathrm{LDA}}(r_s,\eta) - \\ &- \frac{r_s}{3}\frac{\partial \varepsilon_c^{\mathrm{LDA}}(r_s,\eta)}{\partial r_s} - (\eta - sgn\sigma)\frac{\partial \varepsilon_c^{\mathrm{LDA}}(r_s,\eta)}{\partial \eta}, \end{aligned} \tag{A.10}$$

where $sgn\sigma = \pm 1$ for $\sigma =\uparrow$ and $\sigma =\downarrow$, respectively. Using (A.7), for the derivatives we get

$$\begin{aligned}\frac{\partial \varepsilon_c^{\mathrm{LDA}}(r_s,\eta)}{\partial r_s} &= \frac{\partial \varepsilon_c(r_s,0)}{\partial r_s}\left[1 - f(\eta)\eta^4\right] + \\ &+ \frac{\partial \varepsilon_c(r_s,1)}{\partial r_s} f(\eta)\eta^4 + \frac{d\alpha_c(r_s)}{dr_s}\frac{f(\eta)}{f''(0)}(1-\eta^4), \end{aligned} \tag{A.11}$$

and

$$\begin{aligned}\frac{\partial \varepsilon_c^{\mathrm{LDA}}(r_s,\eta)}{\partial \eta} &= 4\eta^3 f(\eta)\left\{\varepsilon_c(r_s,1) - \varepsilon_c(r_s,0) - \frac{\alpha_c(r_s)}{f''(0)})\right\} + \\ &+ f'(\eta)\left\{\eta^4\varepsilon_c(r_s,1) - \eta^4\varepsilon_c(r_s,0) + (1-\eta^4)\frac{\alpha_c(r_s)}{f''(0)}\right\}, \end{aligned} \tag{A.12}$$

where

$$f'(\eta) = \frac{4\left[(1+\eta)^{1/3} - (1-\eta)^{1/3}\right]}{3(2^{4/3}-2)}. \tag{A.13}$$

The r_s derivatives of $\varepsilon_c(r_s,0), \varepsilon_c(r_s,1)$ and $\alpha_c(r_s)$ are obtained from (A.9) as

$$\frac{\partial G}{\partial r_s} = -2A\alpha_1 \ln\left(1 + 1/Q_1\right) - \frac{Q_0 Q_1'}{Q_1^2 + Q_1}, \tag{A.14}$$

with

$$\begin{aligned} Q_0 &= -2A(1+\alpha_1 r_s), \\ Q_1 &= 2A\left(\beta_1 r_s^{1/2} + \beta_2 r_s + \beta_3 r_s^{3/2} + \beta_4 r_s^2\right), && \text{(A.15)} \\ Q_1' &= A\left(\beta_1 r_s^{-1/2} + 2\beta_2 + 3\beta_3 r_s^{1/2} + 4\beta_4 r_s\right). && \text{(A.16)} \end{aligned}$$

A.2 Generalized Gradient Approximation

Within the Generalized Gradient approximation by Perdew, Burke, and Ernzerhof [15], the exchange energy per electron is

$$\varepsilon_x^{\rm GGA}(n,s) = \varepsilon_x^{\rm LDA}(n) F_x^{\rm GGA}(s), \tag{A.17}$$

where

$$s = \frac{|\nabla n|}{2k_F n} \tag{A.18}$$

is the dimensionless or reduced density gradient, $k_F = (3\pi^2 n)^{1/3}$, and

$$F_x^{\rm GGA}(s) = 1 + \kappa - \frac{\kappa}{1+\mu s^2/\kappa}, \tag{A.19}$$

with $\kappa = 0.804$ and $\mu = 0.21951$. The spin polarized GGA exchange energy is obtained from (A.17) and (A.2).

The GGA correlation energy per electron is given as [8, 9, 16]

$$\varepsilon_c^{\rm GGA}(r_s,\eta) = \varepsilon_c^{\rm LDA}(r_s,\eta) + H(r_s,\eta,t), \tag{A.20}$$

where

$$t = \frac{|\nabla n|}{2k_s g(\eta) n}, \tag{A.21}$$

$k_s = (4k_F/\pi)^{1/2}$,

$$g(\eta) = \frac{(1+\eta)^{2/3} + (1-\eta)^{2/3}}{2}, \tag{A.22}$$

and η is defined in (A.6). The function $H(r_s,\eta,t)$ is given by

$$H(r_s,\eta,t) = \gamma g(\eta)^3 \ln\left\{1 + \frac{\beta}{\gamma} t^2 \left[\frac{1+\xi t^2}{1+\xi t^2 + \xi^2 t^4}\right]\right\}, \tag{A.23}$$

where

$$\xi = \frac{\beta}{\gamma} \frac{1}{\exp\left[-\varepsilon_c^{\rm LDA}(n)/2\gamma g(\eta)^3\right] - 1}, \tag{A.24}$$

$\gamma = 0.031091$ and $\beta = 0.066725$.

A.3 Local Airy Gas Approximation

The concept of the edge electron gas was put forward by Kohn and Mattsson [18] as an appropriate basis for the treatment of systems with edge surfaces outside of which all Kohn–Sham orbitals decay exponentially. Its simplest realization, the Airy gas model, is based on the linear potential approximation. Vitos and co-workers [17] used this model to construct a gradient level exchange-energy functional. The so derived exchange functional, that we refer to as the local Airy gas (LAG) functional, has a number of advantages over previous GGA functionals: (a) it explicitly includes the properties of the edge region where much interesting physics occurs, (b) its accuracy may be systematically improved by including higher-order expansions of the effective potential of the model system, and (c) it is as simple and well-defined as that of the standard LDA, *i.e.*, it has no adjustable parameters.

The LAG exchange energy per electron may be written in the form

$$\varepsilon_x^{\mathrm{LAG}}(n,s) = \varepsilon_x^{\mathrm{LDA}}(n) F_x^{\mathrm{LAG}}(s(\zeta)), \tag{A.25}$$

where the enhancement function is given by the following expression

$$\begin{aligned} F_x^{\mathrm{LAG}}(s(\zeta)) = \frac{2}{3\pi}\left(\frac{\pi}{3}\right)^{1/3} \frac{1}{n^{4/3}(\zeta)} \int_{-\infty}^{\infty}\int_0^{\infty}\int_0^{\infty} Ai(\zeta+\epsilon)Ai(\zeta'+\epsilon) \\ \times \quad Ai(\zeta+\epsilon')Ai(\zeta'+\epsilon')\frac{g(\sqrt{\epsilon}|\zeta'-\zeta|, \sqrt{\epsilon'}|\zeta'-\zeta|)}{|\zeta'-\zeta|^3} d\zeta'\, d\epsilon\, d\epsilon'. \end{aligned} \tag{A.26}$$

The dimensionless parameter ζ is obtained from the scaled gradient $s(\zeta)$

$$s(\zeta) \equiv \frac{n'(\zeta)}{2(3\pi^2)^{1/3} n^{4/3}(\zeta)}, \tag{A.27}$$

where the Airy gas density is

$$n(\zeta) = \frac{1}{2\pi}\int_0^{\infty} Ai^2(\zeta+\zeta')\, \zeta' d\zeta'. \tag{A.28}$$

In (A.26) and (A.28), Ai denotes the Airy function [102].

The scaled gradient is conserved when going from the real electron gas to the Airy gas model [331] and therefore the enhancement function $F_x^{\mathrm{LAG}}(s)$ can be parametrized. A modified Becke form [332],

$$F_x^{LAG}(s) = 1 + \beta\frac{s^{\alpha}}{(1+\gamma s^{\alpha})^{\delta}}, \tag{A.29}$$

for $\alpha = 2.626712, \beta = 0.041106, \gamma = 0.092070$, and $\delta = 0.657946$, reproduces the exact result (A.26) for $0 < s < 20$ with an average deviation less than 0.3%. This expression is used to obtain the exchange energy density of the real electron gas from the local, scaled gradient s.

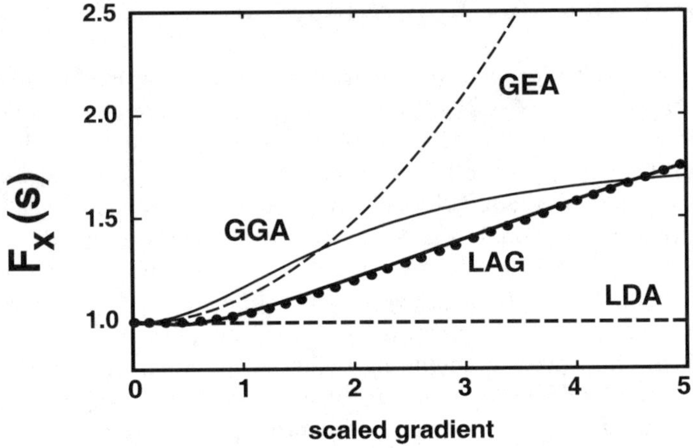

Fig. A.1. The exchange energy enhancement function (A.26) and the parametrized form (A.29), the latter indicated by dots, of the local Airy gas (LAG) compared to those of the local density approximation (LDA), the the generalized gradient approximation (GGA) from (A.19), and the second order gradient expansion (GEA) $F_x^{\mathrm{GEA}}(s) = 1 + 0.1234s^2$ [3].

The LAG enhancement functions calculated from (A.26) and (A.29) are compared to other gradient level enhancement functions in Figure A.1. It follows from the figure that the LAG approximation in the low-gradient limit correctly reduces to the LDA limit. Furthermore, at small gradients, $F_x^{\mathrm{LAG}}(s)$ rises more slowly than both the GGA function (A.19) and the second order Gradient Expansion function [3].

Most of the currently applied approximate density functionals are based on error cancelations between the exchange and correlation energies [157, 333]. Mathematically, this means that for physically interesting densities the total enhancement function over the local exchange energy, *viz.*

$$F_{xc}(s) \equiv \{\varepsilon_x([n], \mathbf{r}) + \varepsilon_c([n], \mathbf{r})\} / \varepsilon_x^{\mathrm{LDA}}(n), \tag{A.30}$$

has a negligible slope up to $s \approx 1$. Combining the LAG exchange functional with the LDA correlation scheme (A.7), the total LAG enhancement function becomes

$$F_{xc}^{\mathrm{LAG}}(s) = F_x^{\mathrm{LAG}}(s) + \varepsilon_c^{\mathrm{LDA}}(n)/\varepsilon_x^{\mathrm{LDA}}(n). \tag{A.31}$$

Thus, the $F_{xc}^{\mathrm{LAG}}(s)$ is determined only by the LAG enhancement function (A.29), which, for $s < 1$, is a slowly increasing function of s (Figure A.1). Therefore, the LAG exchange plus LDA correlation scheme preserves the excellent cancelation properties of the LDA and GGA.

Appendix B

Mathematical Relations

B.1 Real Harmonics

For the spherical harmonics we use the convention from Jackson [87]

$$\bar{Y}_{lm}(\hat{r}) = \left[\frac{(2l+1)(l-m)!}{4\pi(l+m)!}\right]^{1/2} P_l^m(\cos\theta)e^{im\varphi}, \quad \text{(B.1)}$$

where $\hat{r} \equiv (\theta, \varphi)$. The associated Legendre functions are defined for $|m| \leq l$ as

$$P_l^m(x) = \frac{(-1)^m}{2^l l!}(1-x^2)^{m/2}\frac{d^{l+m}}{dx^{l+m}}(x^2-1)^l. \quad \text{(B.2)}$$

The spherical harmonics form a complete orthogonal basis on the spherical surface. They satisfy the normalization and orthogonality condition

$$\int_0^\pi \sin\theta d\theta \int_0^{2\pi} d\varphi \bar{Y}^*_{l'm'}(\theta,\varphi)\bar{Y}_{lm}(\theta,\varphi) = \delta_{l'l}\delta_{m'm}. \quad \text{(B.3)}$$

The integral of three spherical harmonics are the Gaunt numbers

$$G_{lm,l'm',l''m''} = \int_0^\pi \sin\theta d\theta \int_0^{2\pi} d\varphi \bar{Y}_{lm}(\theta,\varphi)\bar{Y}_{l'm'}(\theta,\varphi)\bar{Y}^*_{l''m''}(\theta,\varphi), \quad \text{(B.4)}$$

which are calculated from the usual Clebsch–Gordan or Wigner coefficients. The reader is referred to general mathematical textbooks for further details.

The real harmonics are introduced from the spherical harmonics as

$$Y_{lm}(\hat{r}) = \bar{Y}_{lm}(\hat{r}), \quad \text{(B.5)}$$

for $m = 0$,

$$Y_{lm}(\hat{r}) = \frac{(-1)^m}{i\sqrt{2}} \left\{ \bar{Y}_{lm}(\hat{r}) - \bar{Y}^*_{lm}(\hat{r}) \right\} = \frac{(-1)^m}{\sqrt{2}} 2\mathcal{I}m \left\{ \bar{Y}_{lm}(\hat{r}) \right\}, \qquad \text{(B.6)}$$

for $m < 0$, and

$$Y_{lm}(\hat{r}) = \frac{(-1)^m}{\sqrt{2}} \left\{ \bar{Y}_{lm}(\hat{r}) + \bar{Y}^*_{lm}(\hat{r}) \right\} = \frac{(-1)^m}{\sqrt{2}} 2\mathcal{R}e \left\{ \bar{Y}_{lm}(\hat{r}) \right\}, \qquad \text{(B.7)}$$

for $m > 0$, respectively. These relations may be written in the following condensed form

$$Y_{lm}(\hat{r}) = \sum_{m'} a_{mm'} \bar{Y}_{lm'}(\hat{r}), \qquad \text{(B.8)}$$

where the elements of the transformation matrix a are

$$a_{mm'} = \frac{1}{\sqrt{2}} \begin{cases} i & \text{for } m < 0 \text{ and } m' < 0 \\ -i(-1)^m & \text{for } m < 0 \text{ and } m' > 0 \\ 1 & \text{for } m > 0 \text{ and } m' < 0 \\ (-1)^m & \text{for } m > 0 \text{ and } m' > 0 \\ \sqrt{2} & \text{for } m = 0 \text{ and } m' = 0 \end{cases} \qquad \text{(B.9)}$$

for $|m| = |m'|$ and zero otherwise. After differentiating (B.2), the real harmonics can be recast into

$$Y_{lm}(\hat{r}) = \frac{1}{\sqrt{4\pi}} y_m(\varphi) \sin^{|m|} \varphi \sum_{k \geq (l+|m|)/2}^{l} H_{lk}^{|m|} \cos^{2k-l-|m|} \varphi, \qquad \text{(B.10)}$$

where $y_m(\varphi) = \sin(|m|\varphi)$ for $m < 0$ and $y_m(\varphi) = \cos(|m|\varphi)$ for $m \geq 0$, and

$$H_{lk}^{|m|} \equiv \frac{(-1)^{|m|+k+l}}{2^l l!} \left[(2l+1)(2-\delta_{m0})(l-|m|)!(l+|m|)! \right]^{1/2} \times \frac{l!(2k)!}{k!(l+|m|)!(k-l)!(2k-l-|m|)!}. \qquad \text{(B.11)}$$

The real Gaunt numbers $C_{lm,l'm',l''m''}$ can be obtained from the spherical harmonics Gaunt numbers using the above transformation matrix, *viz.*

$$C_{lm,l'm',l''m''} = \sum_{\mu'',\mu',\mu} a_{m\mu} a_{m'\mu'} a^*_{m''\mu''} G_{l\mu,l'\mu',l''\mu''}. \qquad \text{(B.12)}$$

The real Gaunt numbers are nonzero for $l'' = |l'-l|, |l'-l|+2, ..., l'+l$.

B.2 Rotation Matrix

Upon finite rotation with Euler angles φ, θ, γ the original spherical coordinates θ and ϕ transform into θ' and ϕ'. The corresponding spherical harmonics transform according to the scheme

$$\bar{Y}_{lm}(\hat{r}) = \sum_{m'} \bar{Y}_{lm'}(\hat{r}) \bar{\mathcal{D}}^l_{m'm}(\alpha, \beta, \gamma), \tag{B.13}$$

where

$$\bar{\mathcal{D}}^l_{m'm}(\alpha, \beta, \gamma) \equiv e^{im'\alpha} \bar{d}^l_{m'm}(\beta) e^{im\gamma} \tag{B.14}$$

are the matrix elements of finite rotation for Euler angles φ, θ, γ and quantum numbers l, m, m' [334]. The quantities $\bar{d}^l_{m'm}(\beta)$ may be expressed in terms of the Jacobi polynomial [102] and have the following explicit expression:

$$\begin{aligned} \bar{d}^l_{m'm}(\beta) &= \left[\frac{(l+m')!(l-m')!}{(l+m)!(l-m)!)}\right]^{1/2} \\ &\times \sum_{\mu} \frac{(l+m)!}{(m+m'+\mu)!(l-m'-\mu)!} \frac{(l-m)!}{(l-m-\mu)!\mu!} \\ &\times (-1)^{l-m'-\mu} (\cos\beta/2)^{2\mu+m'+m} (\sin\beta/2)^{2l-2\mu-m'-m} . \end{aligned} \tag{B.15}$$

The matrix elements of finite rotation for real harmonics, defined by

$$Y_{lm}(\hat{r}) = \sum_{m'} Y_{lm'}(\hat{r}) \mathcal{D}^l_{m'm}(\alpha, \beta, \gamma), \tag{B.16}$$

are obtained from $\bar{\mathcal{D}}^l_{m'm}(\alpha, \beta, \gamma)$ using Equation (B.8) and (B.13) and they are given by

$$\mathcal{D}^l_{m'm}(\alpha, \beta, \gamma) = \sum_{\mu\mu'} a_{m\mu} \bar{\mathcal{D}}^l_{\mu'\mu}(\alpha, \beta, \gamma) a^{-1}_{\mu'm'}, \tag{B.17}$$

where a^{-1} is the inverse of the transformation matrix a.

B.3 Bessel and Neumann Functions

The spherical Bessel functions of the first and second kind (often referred to as the Neumann functions) [102] can be represented by the following series expansion

$$J_l(z) = \frac{z^2}{(2l+1)!!}[1 + \mathcal{O}(z^2)] \tag{B.18}$$

and

$$N_l(z) = -\frac{(2l-1)!!}{z^{l+1}}[1+\mathcal{O}(z^2)], \tag{B.19}$$

where $(2l+1)!! \equiv 1 \cdot 3 \cdot 5 ... (2l+1)$. These functions satisfy the following differentiation formulas

$$\begin{aligned} \frac{1}{z}\frac{d}{dz}\left[z^{l+1} f_l(z)\right] &= z^l f_{l-1}(z), \\ \frac{1}{z}\frac{d}{dz}\left[z^{-l} f_l(z)\right] &= -z^{-l-1} f_{l+1}(z), \end{aligned} \tag{B.20}$$

and the recurrence relation

$$f_{l-1}(z) + f_{l+1}(z) = \frac{2l+1}{z} f_l(z). \tag{B.21}$$

For the EMTO basis set in Section 2, for the Bessel and Neumann functions we use the following convention,

$$j_l(\kappa^2, r) = \left(\frac{r}{w}\right)^l \tilde{j}_l(\kappa r) \tag{B.22}$$

and

$$n_l(\kappa^2, r) = \left(\frac{w}{r}\right)^{l+1} \tilde{n}_l(\kappa r), \tag{B.23}$$

where w is the average Wigner–Seitz radius, and the (κr) dependent functions are defined as

$$\tilde{j}_l(\kappa r) \equiv J_l(\kappa r)\frac{(2l-1)!!}{2(\kappa r)^l} = \frac{1}{2(2l+1)}[1+\mathcal{O}\left((\kappa r)^2\right)] \tag{B.24}$$

and

$$\tilde{n}_l(\kappa r) \equiv -N_l(\kappa r)\frac{(\kappa r)^{l+1}}{(2l-1)!!} = 1+\mathcal{O}\left((\kappa r)^2\right). \tag{B.25}$$

With this convention, in the $\kappa \to 0$ limit, we have

$$j_l(\kappa^2, r) \to \left(\frac{r}{w}\right)^l \frac{1}{2(2l+1)} \tag{B.26}$$

and

$$n_l(\kappa^2, r) \to \left(\frac{w}{r}\right)^{l+1}, \tag{B.27}$$

i.e., we recover the functions used within the traditional ASA methods [39]. The energy dependence of the EMTO basis functions is expressed in terms of the dimensionless parameter $\omega \equiv (\kappa w)^2$. Using the differentiation formulas

(B.20) and the relation $\frac{d}{d(\kappa r)} = 2(\kappa r)\left(\frac{w}{r}\right)^2 \frac{d}{d\omega}$, for the ω derivatives of the Bessel and Neumann functions we obtain

$$\frac{\partial j_l(\kappa^2, r)}{\partial(\kappa w)^2} = -\frac{1}{2(2l+1)}\left(\frac{r}{w}\right) j_{l+1}(\kappa^2, r), \tag{B.28}$$

and

$$\frac{\partial n_l(\kappa^2, r)}{\partial(\kappa w)^2} = \frac{1}{2(2l-1)}\left(\frac{r}{w}\right) n_{l-1}(\kappa^2, r). \tag{B.29}$$

These expression may be used to established the higher order ω derivatives of $j_l(\kappa^2, r)$ and $n_l(\kappa^2, r)$, *e.g.*,

$$\frac{\partial^2 n_l(\kappa^2, r)}{\partial(\kappa w)^2 \partial(\kappa w)^2} = \frac{1}{4(2l-1)(2l-3)}\left(\frac{r}{w}\right)^2 n_{l-2}(\kappa^2, r). \tag{B.30}$$

Using again (B.20), the recurrence relation (B.21) and the relation $\frac{d}{d(\kappa r)} = \frac{1}{\kappa}\frac{d}{dr}$, the radial derivatives become

$$\frac{\partial j_l(\kappa^2, r)}{\partial r} = -\frac{l+1}{r} j_l(\kappa^2, r) + \frac{2l-1}{w} j_{l-1}(\kappa^2, r), \tag{B.31}$$

and

$$\frac{\partial n_l(\kappa^2, r)}{\partial r} = \frac{l}{r} n_l(\kappa^2, r) - \frac{2l+1}{w} n_{l+1}(\kappa^2, r). \tag{B.32}$$

Finally, from the differentiation relations (B.20), it is easy to show that the present Bessel and Neumann functions satisfy the Wronskian

$$\mathcal{W}_r\{n_l, j_l\} \equiv r^2 \left[n_l(\kappa^2, r) \frac{\partial j_l(\kappa^2, r)}{\partial r} - \frac{\partial n_l(\kappa^2, r)}{\partial r} j_l(\kappa^2, r) \right] = \frac{w}{2}. \tag{B.33}$$

B.4 Residue Theorem

Let $f(z)$ be a complex function of z and analytic in the whole complex plane, except $z_1, z_2, ..., z_n$, where $f(z)$ is singular. If $f(z)$ is analytic on a closed curve $\mathcal{C}$, then

$$\oint_{\mathcal{C}} f(z)dz = 2\pi i \sum_{i=1,n}{}' Res(f, z_i), \tag{B.34}$$

where the prime means that the summation only includes poles inside $\mathcal{C}$. $Res(f, z_i)$ denotes the -1th order expansion coefficient of $f(z)$ around z_i, *i.e.* if for z near z_i we have

$$f(z) \approx ... + \frac{b_{-2}}{(z-z_i)^2} + \frac{b_{-1}}{(z-z_i)} + b_0 + b_1(z-z_i) + b_2(z-z_i)^2 + ..., \quad \text{(B.35)}$$

then $Res(f, z_i) = b_{-1}$.

As an example, we consider an analytic function of the following form $f(z) = 1/h(z)$. Here the analytic functional $h(z)$ vanishes for the real energies $\epsilon_1, \epsilon_2, ..., \epsilon_n$ and has nonzero derivatives for all ϵ_i, *i.e.* $h(\epsilon_i) = 0$ and $h'(\epsilon_i) \neq 0$ for $i = 1, 2, ..., n$. Expanding $h(z)$ in the vicinity of its roots, *viz.*

$$h(z) = (z - \epsilon_i)h'(\epsilon_i) + ... \text{ for } z \approx \epsilon_i, \quad \text{(B.36)}$$

we obtain

$$Res(f, \epsilon_i) = 1/h'(\epsilon_i). \quad \text{(B.37)}$$

Therefore, the integral of $f(z)$ on a complex contour $\mathcal{C}(\epsilon)$ that includes all ϵ_i $(i = 1, 2, ..., n)$ becomes

$$\oint_{\mathcal{C}(\epsilon)} f(z)dz = 2\pi i \sum_{i=1}^{n} \frac{1}{h'(\epsilon_i)}. \quad \text{(B.38)}$$

We introduce a new function $g(z)$ as

$$g(z) \equiv f(z)h'(z) = \frac{h'(z)}{h(z)}. \quad \text{(B.39)}$$

Since the residuum of $g(z)$ around ϵ_i is

$$Res(g, \epsilon_i) = h'(\epsilon_i)/h'(\epsilon_i) = 1, \quad \text{(B.40)}$$

the integral of $g(z)$ on the complex contour $\mathcal{C}(\epsilon)$ will give the total number of poles for $f(z)$ times $2\pi i$, *i.e.*

$$\frac{1}{2\pi i} \oint_{\mathcal{C}(\epsilon)} g(z)dz = \sum_{i=1}^{n} 1 = n. \quad \text{(B.41)}$$

This relation is used in Sections 2.1.5 and 2.2 to count the single-electron states below the Fermi level.

In the EMTO method the complex contour $\mathcal{C}(\epsilon)$ is a semi-circular or semi-elliptic contour circumscribing the valence states below the Fermi level.

Appendix C

Self-consistent Total Energy Calculations

In this section we shall list the most important numerical details of the EMTO calculations presented in Chapters 7, 8 and 9.

C.1 Numerical Details for Chapter 7

In EMTO calculation of the equilibrium bulk properties presented in Sections 7.1.1 and 7.2.1, the basis set included s, p, d, f orbitals, *i.e.* $l_{max} = 3$ was used (Section 2.1.2). For the slope matrix we used the two-center Taylor expansion (Section 3.4.3) with $n = m = 6$ with $\omega_0 = 0$ and $\omega_1 = -(20 - 40)$ depending on the system. The kink-cancelation equation (Section 2.1.3) was solved for the valence electrons, and the rest of the states, including the high-lying core states, were recalculated after each iteration. This approximation is often named as the soft-core approximation. The Green function was calculated for 16 complex energy points distributed exponentially on a semi-circular contour. The convergency criterion for the total energy and Fermi level was set to 10^{-7} Ry and 10^{-10} Ry, respectively.

For non-metallic solids with A4, B1 and B3 structures (Section 7.2), two empty spheres were included to reduce the overlap error (Equation (2.55)). In $MgSiO_3$ and $ScAlO_3$ perovskites, the overlap between the atomic spheres is $\sim 33\%$, and the muffin-tin discontinuities vary between -0.13 Ry and 0.11 Ry. In order to reduce the overlap error, the radii of the potential spheres around the oxygen atoms were set to $s_O \approx 0.8 w_O$, where w_O is the atomic radius for oxygen. With this choice, the largest linear overlap was decreased to $\sim$19$-$20%, and the error in the kinetic energy to ~ 10 μRy.

The self-consistent calculations were performed using the LDA effective potential and the total energies were computed within the LDA, GGA and LAG approximations. The shape functions were calculated for $l^s_{max} = 30$ (Section 4.1) on a linear mesh between the inscribed and circumscribed containing 120$-$150 mesh points. The equilibrium volume and the bulk modulus were de-

Table C.1. Parameters used for Brillouin-zone sampling in surface calculations for transition metals: P stands for primitive and C for centered, N_k is the number of k-points in the irreducible part of the Brillouin-zone.

Structure	Surface	2D Brillouin-zone	N_k
fcc	(111)	hexagonal	577
	(100)	square	561
	(110)	rectangular-P	560
bcc	(100)	square	378
	(110)	rectangular-C	352
hcp	(0001)	hexagonal	293

termined from the Morse function (Section 6.1) fitted to the *ab initio* total energies calculated for five different atomic volumes.

In surface calculations, the surfaces were modeled by repeated slabs containing N_a atomic layers and separated by a vacuum layer having a width equal to the width of N_v bulk-type atomic layers. The values for N_a and N_v used for the low index surfaces of the *hcp*, *bcc* and *fcc* structures (Section 7.1.3) are given in Table 6.2. In Table C.1, we list the number of inequivalent k-points used to map 2D Brillouin-zones. For surface stress calculation, we used five in-plane distortions $\epsilon = 0.0, \pm 0.01$ and ± 0.02 (Equation (6.103)).

The equation of state and elastic properties of solid He were computed using s, p, d orbitals. The Green function was calculated for 32 complex energy points distributed exponentially on a semicircular contour enclosing the He $1s^2$ states.

C.2 Numerical Details for Chapter 8

In impurity calculations (Section 8.1), the Brillouin zone of the 16-atoms supercell was sampled using ~ 100 k-points in the irreducible wedge.

For elastic constant calculations for cubic Al-based alloys, we used about 15000 uniformly distributed k-points in the irreducible wedge of the orthorhombic and monoclinic Brillouin zones. However, for small concentrations (below $\sim 5\%$) a significantly higher number of k-points ($\sim 10^5$) was needed to obtain well converged elastic constants.

For hexagonal AgZn alloys, in the irreducible wedge of the Brillouin zones $2500-7000$ k-points were used, depending on the particular crystal symmetry.

C.3 Numerical Details for Chapter 9

For FeCrNi alloys, the single-electron equations were solved within the scalar-relativistic and frozen-core approximations, *i.e.* the core states were fixed to the initial atomic states. The Green function was calculated for 16 complex energy points, the EMTO basis set included s, p, d, f orbitals, and in the one-center expansion of the full charge density we used $l^h_{\max} = 10$ (Section 5.2.2). The conventional Madelung energy was calculated for $l^m_{\max} = 8$. In the irreducible wedge of the Brillouin zones we used 1000−1400 k-points, depending on the particular distortion.

For stacking fault energies, the Green function for the valence states was calculated for 52 complex energy points. In the irreducible wedge of the *fcc*, *hcp* and *dhcp* Brillouin zones we used 1000−1500 uniformly distributed k-points.

The EMTO calculations for the elastic constants of FeCrNi alloys were carried out for quaternary alloys, *i.e.* taking into account the disordered local magnetic moments only on Fe sites. In the stacking fault energy calculations, on the other hand, the local magnetic moments on Cr and Ni sites were considered as well.

References

1. Hohenberg, P., Kohn, W.: Phys. Rev. **136B**, 864 (1964)
2. Kohn, W., Sham, L. J.: Phys. Rev. **140A**, 1133 (1965)
3. Dreizler, R. M., Gross, E. K. U.: Density Functional Theory. Springer-Verlag, Berlin Heidelberg (1990)
4. Perdew, J. P., Zunger, A.: Phys. Rev. B **23**, 5048 (1981)
5. Vosko, S.H., Wilk, L., Nusair, M.: Can. J. Phys. **58**, 1200 (1980)
6. Ceperley, D. M., Alder, B. J.: Phys. Rev. Lett. **45**, 566 (1980)
7. Perdew, J., Wang, Y.: Phys. Rev. B **45**, 13244 (1992)
8. Perdew, J. P.: In: Ziesche, P., Eschrig, H. (eds) Electronic Structure of Solids. Akademie Verlag, Berlin, **11** (1991)
9. Perdew, J. P., Chevary, J. A., Vosko, S. H., Jackson, K. A., Pederson, M. R., Singh, D. J., Fiolhais C.: Phys. Rev. B **46**, 6671 (1992)
10. Wang, C. S., Klein, B. M., Krakauer, H.: Phys. Rev. Lett. **54**, 1852 (1985)
11. Leung, T. C., Chan, C. T., Harmon, B. N.: Phys. Rev. B **44**, 2923 (1991)
12. Perdew, J.P., Yue, W.: Phys. Rev. B **33**, 8800 (1986)
13. Perdew, J. P.: Phys. Rev. B **33**, 8822 (1986)
14. Becke, A. D.: Phys. Rev. A **38**, 3098 (1988)
15. Perdew, J. P., Burke, K., Ernzerhof, M.: Phys. Rev. Lett. **77**, 3865 (1996)
16. Perdew, J. P., Burke, K., Wang, Y.: Phys. Rev. B **54**, 16533 (1996)
17. Vitos, L., Johansson, B., Kollár, J., Skriver, H. L.: Phys. Rev. B **62**, 10046 (2000)
18. Kohn, W., Mattsson, A.E.: Phys. Rev. Lett. **81**, 3487 (1998)
19. Slater, J. C.: Phys. Rev. **51**, 151 (1937)
20. Andersen, O.K.: Phys. Rev. B **12**, 3060 (1975)
21. Singh, D. J.: In: Planewaves, Pseudopotentials and the LAPW Method. Kluwer Academic Publishers, Boston (1994)
22. Drittler, B., Weinert, M., Zeller, R., Dederichs, P. H.: Solid State Commun. **79**, 31 (1991)
23. Papanikolaou, N, Zeller, R., Dederichs, P. H., Stefanou, N.: Phys. Rev. B **55**, 4157 (1997)
24. Huhne, T., Zecha, C., Ebert, H., Dederichs, P. H. Zeller, R.: Phys. Rev. B **58**, 10236 (1998)
25. Wills, J. M., Cooper, B.: Phys. Rev. B **36**, 389 (1987)
26. Methfessel, M.: Phys. Rev. B **38**, 1537 (1988)

27. Savrasov, S. Y., Savrasov, D. Y.: Phys. Rev. B **54**, 16487 (1996)
28. Phillips, J. C.: Phys. Rev. **112**, 685 (1958)
29. Cohen, M. L., Heine, V.: Solid State Phys. **24**, 37 (1970)
30. Kresse, G., Furthmüller, J.: Phys. Rev. B **55**, 11169 (1996)
31. Hutter, J., Luthi, H. P., Parrinello, M.: Comput. Mater. Sci. **2**, 244 (1994)
32. Bachelet, G. B., Hamann, D. R., Schluter, M.: Phys. Rev. B **26**, 4199 (1982)
33. Hartwigsen, C., Goedecker, S., Hutter, J.: Phys. Rev. B **58**, 3641 (1998)
34. Kiejna, A., Kresse, G., Rogal, J., De Sarkar, A., Reuter, K., Scheffler, M.: Phys. Rev. B **73**, 035404 (2006)
35. Car, R., Parrinello, M.: Phys. Rev. Lett. **55**, 2471 (1985)
36. Korringa, J.: Physica **13**, 392 (1947)
37. Kohn, W., N. Rostoker, N.: Phys. Rev. **94**, 1111 (1954)
38. Szunyogh, L., Újfalussy, B., Weinberger, P., Kollár, J.: Phys. Rev. B **49**, 2721 (1994)
39. Skriver, H. L.: The LMTO Method. Springer-Verlag, Berlin Heidelberg New York Tokyo(1984)
40. Andersen, O. K., Jepsen, O., Sob, M.: Linearized Band Structure Methods. In: Yussouff, M. (ed.) Lecture Notes in Physics: Electronic Band Structure and Its Applications. Springer-Verlag, Berlin (1987)
41. Kübler, J.: In: Theory of Itinerant Electron Magnetism. Clarendon Press, Oxford (2000)
42. Andersen, O. K., Postnikov, A. V., Savrasov, S. Y.: In Butler, W. H, Dederichs, P. H., Gonis, A., Weaver, R. L. (eds) Applications of Multiple Scattering Theory in Materials Science. Materials Research Society, Pittsburgh, PA (1992)
43. Asato, M., Settels, A., Hoshino, T., Asada, T., Blügel, S., Zeller, R., Dederichs, P.H.: Phys. Rev. B **60**, 5202 (1999)
44. Kollár, J., Vitos, L., Skriver, H. L.: From ASA Towards the Full Potential. In: Dreyssé, H. (ed.) Lecture Notes in Physics: Electronic Structure and Physical Properties of Solids. Springer-Verlag, Berlin Heidelberg New York Tokyo (2000)
45. Vitos, L., Kollár, J., Skriver, H. L.: Phys. Rev. B **55**, 13521 (1997)
46. Andersen, O. K., Jepsen, O., Krier, G.: In: Kumar, V., Andersen, O.K., Mookerjee, A. (eds) Lectures on Methods of Electronic Structure Calculations, World Scientific Publishing Co., Singapore, **63** (1994)
47. Andersen, O. K., Arcangeli, C. (unpublished).
48. Vitos, L., Kollár, J., Skriver, H. L.: Phys. Rev. B **49**, 16694 (1994)
49. Kollár, J., Vitos, L., Skriver, H. L.: Phys. Rev. B **49**, 11288 (1994)
50. Vitos, L., Kollár, J., Skriver, H. L.: Phys. Rev. B **55**, 4947 (1997)
51. Weinberger, P.: In: Electron Scattering Theory for Ordered and Disordered Matter. Clarendon Press, Oxford (1990)
52. Gonis, A.: In: van Groesen, E., de Jager, E. M. (eds) Green Functions for Ordered and Disordered Systems. Studies in Mathematical Physics **4**, North Holland (1992)
53. Rickayzen, G.: In: March, N. H. (ed) Techniques of Physics: Green's Functions and Condensed Matter. Academic Press, London (1984)
54. Ledbetter, H. M.: J. Appl. Phys. **44**, 1451 (1973)

55. Ledbetter, H. M.: In: Levy, M., Bass, H. E., Stern, R. R. (eds) Handbook of Elastic Properties of Solids, Liquids, Gases. Academic, San Diego, volume **III**, 313 (2001)
56. Althoff, J. D., Morgan, D., de Fontaine, D., Asta, M. D., Foiles, S. M., Johnson, D. D.: Phys. Rev. B **56**, R5705 (1997)
57. Ravelo, R., Aguilar, J., Baskes, M. I., Angelo, J. E., Fultz, B., Holian, B. L.: Phys. Rev. B **57**, 862 (1998)
58. Goedecker, S.: Rev. Mod. Phys. **71**, 1085 (1999)
59. Abrikosov, I. A., Niklasson, A. M. N., Simak, S. I., Johansson, B., Ruban, A. V., Skriver, H. L.: Phys. Rev. Lett. **76**, 4203 (1996)
60. Stripp, K. F., Kirkwood, J. G.: J. Chem. Phys. **22**, 1579 (1954)
61. Wojtowitcz, P. J., Kirkwood, J. G.: J. Chem. Phys. **33**, 1299 (1960)
62. Bellaiche, L., Vanderbilt, D.: Phys. Rev. B **61**, 7877 (2000)
63. van de Walle, A., Ceder, G.: Rev. Modern Phys. **74**, 11 (2002)
64. Sanchez, J. M., Ducastelle, G., Gratias, D.: Physica A **128**, 334 (1984)
65. de Fontaine, D., Asta, M.: Int. Conf. on Computer-assisted Materials Design and Process Simulation, Tokyo, ISIJ, 272 (1993)
66. Connolly, J. W. D., Williams, A. R.: Phys. Rev. B **27**, 5169 (1983)
67. Zunger, A., Wei, S. H., Ferreira, L. G., Bernard, J. E.: Phys. Rev. Lett **65**, 353 (1990)
68. Soven, P.: Phys. Rev. **156**, 809 (1967)
69. Taylor, D. W.: Phys. Rev. **156**, 1017 (1967)
70. Györffy, B. L.: Phys. Rev. B **5**, 2382 (1972)
71. Abrikosov, I. A., Vekilov, Yu. H., Ruban, A. V.: Soviet Phys., Solid State **32**, 2004 (1990)
72. Abrikosov, I. A., Vekilov, Yu. H., Ruban, A. V.: Phys. Lett. A **154**, 407 (1991)
73. Abrikosov, I. A., Skriver, H. L.: Phys. Rev. B **47**, 16532 (1993)
74. Singh, P. P., Gonis, A.: Phys. Rev. B **49**, 1642 (1994)
75. Temmerman, W. M., Gyorffy, B. L., Stocks, G. M.: J. Phys. F **8**, 2461 (1978)
76. Singh, P. P., de Fontaine, D., Gonis, A.: Phys. Rev. B **44**, 8578 (1991)
77. Antropov, V. P., Harmon, B. N.: Phys. Rev. B **51**, 1918 (1995)
78. Vitos, L., Abrikosov, I. A., Johansson, B.: Phys. Rev. Lett **87**, 156401 (2001)
79. Vitos, L.: Phys. Rev. B **64**, 014107 (2001)
80. Vitos, L., Skriver, H. L., Johansson, B., Kollár, J.: Comput. Mater. Sci. **18**, 24 (2000)
81. Rowlands, D. A., Staunton, J. B., Gyrffy, B. L.: Phys. Rev. B **67**, 115109 (2003)
82. Rowlands, D. A., Staunton, J. B., Gyrffy, B. L., Bruno, E., Ginatempo, B.: Phys. Rev. B **72**, 045101 (2005)
83. Andersen, O.K., Jepsen, O., Glötzel, D.: In: Bassani, F., Fumi, F., Tosi, M.P. (eds) Highlights of Condensed-Matter Theory. North-Holland, New York (1985)
84. Andersen, O. K., Arcangeli, C., Tank, R. W., Saha-Dasgupta, T., Krier, G., Jepsen, O., Dasgupta, I.: In Mater. Res. Soc. Symp. Proc. **491**, 3 (1998)
85. Andersen, O. K., Saha-Dasupta, T.: Phys. Rev. B **62**, R16219 (2000)

86. Andersen, O. K., Saha-Dasgupta, T., Tank, R. W., Arcangeli, C., Jepsen, O., Krier, G.: Developing the MTO Formalism. In: Dreyssé, H. (ed.) Lecture Notes in Physics: Electronic Structure and Physical Properties of Solids. Springer-Verlag, Berlin Heidelberg New York Tokyo (2000)
87. Jackson, J.D.: Classical Electrodynamics. Wiley, New York (1975)
88. Nagamatsu, J., Nakagawa, N., Muranaka, T., Zenitani, Y., Akimitsu, J.: Nature **410**, 63 (2001)
89. Slusky, J. S., Rogado, N., Regan, *et al.*: Nature **410**, 343 (2001)
90. Renker, B., Bohnen, K. B., Heid, R., Ernst, D., Schober, H., Koza, M., *et al.*: Phys. Rev. Lett. **88**, 067001 (2002)
91. Kissavos, A. E., Vitos, L., Abrikosov, I. A.: Phys. Rev. B **75**, 115117 (2007)
92. Rodriguez, O. C., Methfessel, M.: Phys. Rev. B **45**, 90 (1992)
93. Zabloudil, J., Hammerling, R., Szunyogh, L., Weinberger, P.: Phys. Rev. B **73**, 115410 (2006)
94. Vitos, L., Kollár, J.: Phys. Rev. B **51**, 4074 (1995)
95. Gonis, A., Sowa, E. C., Sterne, P. A.: Phys. Rev. Lett. **66**, 2207 (1991)
96. Andersen, O. K., Wolley, R. G.: Mol. Phys. **26**, 905 (1973)
97. van W. Morgan, J.: J. Phys. C **10**, 1181 (1977)
98. Stefanou, N., Akai, H., Zeller, R.: Comput. Phys. Commun. **60**, 231 (1990)
99. Schadler, G. H.: Phys. Rev. B **45**, 11314 (1992)
100. Butler, W. H., Gonis A., Zhang, X.- G.: Phys. Rev. B **48**, 2118 (1993)
101. Zhang, X.-G., Butler, W. H., MacLaren, J. M., van Ek, J.: Phys. Rev. B **49**, 13383 (1994)
102. Abramowitz, M., Stegun, I.A. (eds) Handbook of Mathematical Functions, Dover Publications, Inc., New York (1970)
103. Ham, F. S., Segall, B.: Phys. Rev. **124**, 1786 (1961)
104. Faulkner, J. S., Davis, H. L., Joy, H. W.: Phys. Rev. **161**, 656 (1967)
105. Vitos, L., Korzhavyi, P. A., Johansson, B.: Phys. Rev. Lett. **88**, 155501(4) (2002)
106. Faulkner, J. S., Stocks, G. M.: Phys. Rev. B **21**, 3222 (1980)
107. Faulkner, J. S.: Prog. Mater. Sci. **27**, 1 (1982)
108. Johnson, D. D., Nicholson, D. M., Pinski, F. J., Györffy, B. L., Stocks, G. M.: Phys. Rev. Lett. **56**, 2088 (1986)
109. Weinberger, P., Drchal, V., Szunyogh, L., Fritscher, J., Bennet, B. I.: Phys. Rev. B **49**, 13366 (1994)
110. Akai, H.: J. Phys. Condens. Matter **1**, 8045 (1989)
111. Abrikosov, I. A., Vekilov, Yu. H., Korzhavyi, P. A., Ruban, A. V., Shilkrot, L. E.: Solid State Commun. **83**, 867 (1992)
112. Simak, S. I., Ruban, A. V., Vekilov, Yu. H. Solid State Commun. **87**, 393 (1993)
113. Singh, P. P., Gonis, A., Turchi, P. E. A.: Phys. Rev. Lett. **71**, 1605 (1993)
114. Johnson, D. D., Nicholson, D. M., Pinski, F. J., Györffy, B. L., Stocks, G. M.: Phys. Rev. B **41**, 9701 (1990)
115. Magyari-Köpe, B., Grimvall, G., Vitos, L.: Phys. Rev. B **66**, 064210 (2002)
116. Magyari-Köpe, B., Grimvall, G., Vitos, L.: Phys. Rev. B **66**, 179902 (2002).
117. Ruban, A. V., Abrikosov, I. A., Skriver, H. L.: Phys. Rev. B **51**, 12958 (1995)

118. Györffy, B. L., Stocks, G. M.: In: Phariseau, P., *et al.* (eds) Electrons in Disordered Matals and Metallic Surfaces. Plenum, New York (1987)
119. James, P., Abrikosov, I. A., Eriksson, O., Johansson, B.: In: Gonis, A., Meike, A., Turchi, P. E. A. (eds) Properties of Complex Inorganic Solids. Plenum, New York, p. 57 (1997)
120. Abrikosov, I. A., Johansson, B.: Phys. Rev. B **57**, 14164 (1998)
121. Korzhavyi, P. A., Ruban, A. V., Abrikosov, I. A., Skriver, H. L.: Phys. Rev. B **51**, 5773 (1995)
122. Ruban, A. V., Skriver, H. L.: Phys. Rev. B **66** 024201 (2002)
123. Moruzzi, V. L., Janak, J. F., Schwarz, K.: Phys. Rev. B **37**, 790 (1988)
124. Murnaghan, F.D.: Proc. Nat. Acad. Sci. **30**, 244 (1944)
125. Fu, C.-L., Ho, K.-M.: Phys. Rev. B, 5480 (1983)
126. Birch, F.: Phys. Rev. **71**, 809 (1947)
127. Oganov, A. R., Dorogokupets, P. I.: Phys. Rev. B **67**, 224110 (2003)
128. Grimvall, G.: Thermophysical Properties of Materials. North-Holland, Amsterdam (1999)
129. Young, D. A.: Phase Diagrams of the Elements. University of California Press, Berkeley (1991)
130. Every, A. G.: Phys. Rev. B **22**, 1746 (1980)
131. Zener, C,: Elasticity and Anelasticity of Metals. University of Chicago Press, Chicago (1948)
132. Steinle-Neumann, G., Stixrude, L. Cohen R. E.: Phys. Rev. B **60**, 791 (1999)
133. Fast, L., Wills, J.M., Johansson, B., Eriksson, O.: Phys. Rev. B **51**, 17431 (1995)
134. Chung, D. H., Buessem, W. R.: J. Appl. Phys. **38**, 2010 (1967)
135. Hashin, Z., Shtrikman, S.: J. Mech. Phys. Solids **10**, 343 (1962)
136. Hill, R.: Proc. Phys. Soc. A **65**, 349 (1952)
137. Hill, R.: J. Mech. Phys. Solids **11**, 357 (1963)
138. Tyson, W. R., Miller, W.A.: Surf. Sci. **62**, 267 (1977)
139. de Boer, F. R., Boom, R., Mattens, W. C. M., Miedema, A.R., Niessen, A.K.: Cohesion in Metals, North-Holland, Amsterdam, (1988)
140. Métois, J. J., Saúl, A., Müller, P.: Nature Mater. **4**, 238 (2005)
141. Shuttleworth, R.: Proc. Phys. Soc. **A 63**, 445 (1950)
142. Denteneer, P. J. H., van Haeringen, W.: J. Phys.C: Solid State Phys. **20**, L883 (1987)
143. Cheng, C., Needs, R. J., Heine, V.: J. Phys. C: Solid State Phys. **21**, 1049 (1988)
144. Argon, A. S., Backer, S., McClintock, F. A., Reichenbach, G. S., Orowan, E., Shaw, M. C., Rabinowicz, E. In: McClintock, F. A., Argon, A. S.: Addison-Wesley Series in Metallurgy and Materials. Addison-Wesley Publishing Company, Inc., Ontario (1966)
145. Fleischer, R. L.: Acta Met. **11**, 203 (1963)
146. Labusch, R.: Acta Met. **20**, 917 (1972)
147. Nabarro, F. R. N.: Philos. Mag. **35**, 613 (1977)
148. Lung, C. W., March, N. H.: Mechanical Properties of Metals: Atomistic and Fractal Continuum Approaches. World Scientific Publishing Co. Pte. Ltd. (1999)
149. Clerc, D. G., Ledbetter, H. M.: J. Phys. Chem. Solids **59**, 1071 (1998)

150. Hugosson, H., Jansson, U., Johansson, B., Eriksson, O.: Science **293**, 2423 (2001)
151. Sander, D.: Curr. Opin. Solid State Mater. Sci. **7**, 51 (2003)
152. Mechler, A., Kokavecz, J., Heszler, P., Lal, R.: Appl. Phys. Lett. **82**, 3740 (2003)
153. Heszler, P., Révész, K., Reimann, C. T., Mechler, A., Bor, Z.: Nanotechnology **11**, 37 (2000)
154. Schramm, R. E., Reed, R. P.: Met. Trans. A **6A**, 1345 (1975)
155. Miodownik, A. P.: Calphad **2**, 207 (1978)
156. Ozilins, V., Körling, M.: Phys. Rev. B **48**, 18304 (1993)
157. Kurth, S., Perdew, P. J., Blaha, P.: Int. J. Quantum Chem. **75**, 889 (1999)
158. Villars, P, Calvert, L. D.: Pearsons Handbook of Crystallographic Data for Intermetallic Phases. American Society for Metals, Metals Park, OH 44073 (1989)
159. Akahama, Y., Fujihisa, H., Kawamura, H.: Phys. Rev. Lett. **94**, 195503 (2005)
160. Geiselmann, J.: J. Less-Common Met. **4**, 362 (1962)
161. Chetty, N., Weinert, M., Rahman, T. S., Davenport, J. W.: Phys. Rev. B **52**, 6313 (1995)
162. Turner, D. E., Zhu, Z. Z., Chan, C. T., Ho, K. M.: Phys. Rev. B **55**, 13842 (1997)
163. Hoshino, T., Papanikolaou, N., Zeller, R., Dederichs, P. H., Asato, M., Asada, T., Stefanou, N.: Comput. Mater. Sci. **14**, 56, (1999)
164. Ehrhart, P., Jung, P., Schulta, H., Ullmaier, H.: In: Ullmaier, H. (ed.) Atomic Defects in Metals. Landolt-Bornstein, New Series Group III, Vol. **25** Springer-Verlag, Berlin (1990)
165. Kollár, J., Vitos, L., Osorio-Guillén, J. M., Ahuja, R.: Phys. Rev. B **68**, 245417 (2003)
166. Ibach, H.: Surf. Sci. Rep. **29**, 193 (1997)
167. Kádas, K., Nabi, Z., Kwon, S. K., Vitos, L., Ahuja, R., Johansson, B., Kollár, J.: Surf. Sci. **600**, 395 (2006)
168. Kwon, S.K., Nabi, Z., Kádas, K., Vitos, L., Kollár, J., Johansson, B., Ahuja, R.: Phys. Rev. B **72**, 235423 (2005)
169. Blchl, P. E.: Phys. Rev. B **50**, 17953 (1994)
170. Methfessel, M., Hennig, D., Scheffler, M.: Phys. Rev. B **46**, 4816 (1992)
171. Vitos, L., Ruban, A. V., Skriver, H. L., Kollár, J.: Surf. Sci. **411**, 186 (1998)
172. Martienssen, W., Warlimont, H.: Springer Handbook of Condensed Matter and Materials Data, Springer-Verlag (2005)
173. Heine, V., Marks, L. D.: Surf. Sci. **165**, 65 (1986)
174. Filippetti, A., Fiorentini, V., Stokbro, K., Valente, R., Baroni, S.: Mater. Res. Soc. Symp. Proc. **408**, 457 (1996)
175. Fiorentini, V., Methfessel, M., Scheffler, M.: Phys. Rev. Lett. **71**, 1051 (1993)
176. Feibelman, P.J.: Phys. Rev. B **51**, 17867 (1995)
177. Lamber, R., Wetjen, S., Jaeger, N.I.: Phys. Rev. B **51**, 10968 (1995)
178. Wasserman, H.J., Vermaak, J.S.: Surf. Sci. **22**, 164 (1970)
179. Takemura, K.: J. Appl. Phys. **89**, 662 (2001)
180. Bell, P. M., Mao, H. K.: Year Book - Carnegie Inst. Washington **80**, 404 (1981)

181. Downs, R. T., Zha, C. S., Duffy, T. S., Finger, L. W.: Am. Mineralogist **81**, 51 (1996)
182. Takemura, K., Sato, K., Fujihisa, H., Onoda, M.: Ferroelectrics **305**, 103 (2004)
183. Zha, C. S., Mao, H. K., Hemley, R. J.: Phys. Rev. B **70**, 174107 (2004)
184. Nabi, Z., Vitos, L., Johansson, B., Ahuja, R.: Phys. Rev. B **72**, 172102 (2005)
185. Loubeyre, P., LeToullec, R., Pinceaux, J. P., Mao, H.K., Hu, J., Hemley, R.J.: Phys. Rev. Lett. **71**, 2272 (1993)
186. Franck, J. P., Daniels, W. B.: Phys. Rev. Lett. **44**, 259 (1980)
187. Mao, H. K., Hemley, R. J., Wu, Y., Jephcoat, A. P., Finger, L. W., Zha, C. S., Bassett, W. A.: Phys. Rev. Lett. **60**, 2649 (1988)
188. Söderlind, P., Eriksson, O., Wills, J. M., Boring, A. M.: Phys. Rev. B **48**, 5844 (1993)
189. Landolt-Börnsteind, In: Numerical Data and Functional Relationships in Science and Technology. New Series, Group III, Vol. **1**, Springer-Verlag, Berlin (1966)
190. Knittle, E., Jeanloz, R.: Science **235**, 668 (1987)
191. Kung, J., Rigden, S., Jackson, I.: Phys. Earth Planet. Int., **120**, 299 (2000)
192. Kung, J., Rigden, S. M., Gwanmesia, G.: Phys. Earth Planet. Int., **118**, 65 (2000)
193. Hill, R. J., Jackson, I.: Phys. Chem. Miner. **17**, 89, (1990)
194. Reid, A. F., Ringwood, A. E.: J. Geophys. Res. **80**, 3363, (1975)
195. Ross, N. L.: Phys. Chem. Miner. **25**, 597, (1998)
196. Ross, N. L., Hazen, R. M.: Phys. Chem. Miner. **17**, 228, (1990)
197. Magyari-Köpe, B., Vitos, L., Johansson, B., Kollár, J.: Acta Crystallogr. **B57**, 491 (2001)
198. Magyari-Köpe, B., Vitos, L., Johansson, B., Kollár, J.: Phys. Rev. B **63**, 104111 (2001)
199. Horiuchi, H., Ito, E., Weidner, D. J.: Am. Mineralogist **72**, 357 (1987)
200. Bass, J. D.: Phys. Earth Planet. Int. **36**,145, (1984)
201. Liebermann, R. C., Jones, L. E. A., Ringwood, A. E.: Phys. Earth Planet. Int. **14**, 165, (1977)
202. Stixrude, L., Cohen, R.E.: Nature **364**, 613, (1993)
203. Wentzcovitch, R. M., Martins, J. L., Price, G. D.: Phys. Rev. Lett. **70**, 3947 (1993)
204. Lee, T., Ohmori, K., Shin, C.-S., Cahill, D. G., Petrov, I., Greene, J. E.: Phys. Rev. B **71**, 144106 (2005)
205. Carlson, O. N., Smith, J. F., Nafziger, R. H.: Met. Trans. A **17**, 1647 (1986)
206. Kim, J. O., Achenbach, J. D., Mirkarimi, P. B., Shinn. M., Barnett, S. A.: J. Appl. Phys. **72**, 1805 (1992)
207. Dudiy, S. V., Lundqvist, B. I.: Phys. Rev. B **64**, 045403 (2001)
208. Hartford, J.: Phys. Rev. B **61**, 2221 (2000)
209. Ahuja, R., Eriksson, O., Wills, J. M., Johansson, B.: Phys. Rev. B **53**, 3072 (1996)
210. Vitos, L., Larsson, K., Johansson, B., Hanson, M., Hogmark, S.: Comput. Mat. Sci., **37**, 193 (2006)
211. Marlo, M, Milman, V.: Phys. Rev. B **62**, 2899 (2000)

212. Liu, W., Liu, X., Zheng, W. T., Jiang, Q.: Surf. Sci. **600**, 257 (2006)
213. Gauthier, Y., Joly, Y., Rundgren, J., Johansson, L. I., and Wincott, P.: Phys. Rev. B **42**, 9328 (1990)
214. Korzhavyi, P. A., Ruban, A. V., Simak, S. I., Vekilov, Yu. Kh.: Phys. Rev. B **49**, 14229, (1994)
215. Sluiter, M., de Fontaine, D., Guo, X. Q., Podloucky, R., Freeman, A. J.: Phys. Rev. B **42**, 10460 (1990)
216. Sluiter, M. H. F., Watanabe, Y., de Fontaine, D., Kawazoe, Y.: Phys. Rev. B **53**, 6137 (1996)
217. Vaks, V. G., Zein, N. E.: J. Phys. Condens. Matter **2**, 5919 (1990)
218. Kellington, S. H., Loveridge, D., Titman, J. M.: Br. J. Appl. Phys. (J. Phys. D) Ser. 2, **2**, 1162 (1969)
219. Müller, W., Bubeck, E., Gerold, V.: In: Baker, C., *et al.* (eds) Aluminium-Lithium Alloys. vol. III, London Institute of Metals **435** (1986)
220. Taga, A., Vitos, L., Johansson, B., Grimvall, G.: Phys. Rev. B **71**, 014201 (2005)
221. Mehl, M. J.: Phys. Rev. B **47**, 2493 (1993)
222. Clouet, E., Sanchez, J. M., Sigli, C.: Phys. Rev. B **65**, 094105 (2002)
223. Sinko, G. V., Smirnov, N. A.: J. Phys.: Condens. Matter **14**, 6989 (2002)
224. Yang, S. H., Mehl, M. J., Papaconstantopoulos, D. A.: Phys. Rev. B **57**, R2013 (1998)
225. Kamm, G. N., Alers, G. A.: J. Appl. Phys. **35**, 327 (1964)
226. Axon, H. J., Hume-Rothery, W.: Proc. Roy. Soc. **A193**, 1 (1948)
227. Ruban, A. V., Simak, S. I., Shallcross, S., Skriver, H. L.: Phys. Rev. B **67**, 214302 (2003)
228. Sankaran, K. K., Grant, N. J.: Mater. Sci. Eng. **44**, 213 (1980)
229. Pugh, S. F.: Philos. Mag. **45**, 823 (1954)
230. Gault, C., Dauger, A., Boch, P.: Physica Status Solidi (a) **43**, 625 (1977)
231. King, H. W.: J. Mater. Sci. **1** 79 (1966)
232. Pearson, W.B.: Handbook of Lattice Spacings and Structures of Metals and Alloys. Raynor,G. V. (ed.) Pergamon Press Ltd. (1958)
233. Wang, Y., Stocks, G. M., Shelton, W. M., Nicholson, D. M. C., Temmerman, W. M., Z. Szotek, Z.: Phys. Rev. Lett. **75**, 2867 (1995)
234. Massalski, T.: Binary Alloys Phase Diagrams. ASM, Metals Park, Ohio, 117 (1990)
235. Magyari-Köpe, B., Vitos, L., Grimvall, G.: Phys. Rev. B **70**, 052102 (2004)
236. Ashcroft, N. W., Mermin, N. D.: Solid State Physics. Saunders College Publishing, USA (1976)
237. Khein, A., Singh, D. J., Umrigar, C. J.: Phys. Rev. B **51**, 4105 (1995)
238. Müller, Wang, S. L. W., Zunger, A., Wolverton, C.: Phys. Rev. B **60**, 16448, (1999)
239. Murakami, Y., and Kachi, S.: Jpn. J. Appl. Phys. **13**, 1728, (1974)
240. Every, A. G., McCurdy, A. K.: In Landolt Börnstein tables, New Series III/29a, **1**, Springer Verlag, Berlin (1992)
241. Mehl, M. J., Papaconstantopoulos, D. A.: Phys. Rev. B **54**, 4519, (1996)
242. Bruno, E., Ginatempo, B., Giuliano, E. S., Ruban, A. V., and Vekilov, Yu. K.: Phys. Rep. **249**, 353 (1994)
243. Wills, J. M., Eriksson, O., Söderlind, P., Boring, A. M.: Phys. Rev. Lett. **68** 2802 (1992)

244. Green, B. A. Jr.: Phys. Rev. **144**, 528 (1966)
245. Dubrovinsky, L., *et al.*: Nature **422**, 58 (2003)
246. Vitos, L., Korzhavyi, P. A., Johansson, B.: Nature Mater. **2**, 25 (2003)
247. Massalski, T. B.: J. Phys. Radium **23**, 647, (1962)
248. Matsuo, Y.: J. Phys. Soc. Japan **53**, 1360, (1984)
249. Steinle-Neumann, G., Stixrude, L., Cohen, R. E.: Phys. Rev. B **63**, 054103, (2001)
250. Häussermann, U., Simak, S. I.: Phys. Rev. B **64**, 245114 (2001)
251. Zheng-Johansson, J. X., Eriksson, O. E., Johansson, B.: Phys. Rev. B **59**, 6131 (1999)
252. Singh, D., Papaconstantopoulos, D. A.: Phys. Rev. B **42**, 8885 (1990)
253. Dubrovinskaia, N., *et al.*: Phys. Rev. Lett. **95**, 245502 (2005)
254. Massalski, T. B.: Structure and stability of alloys. In: Cahn, R. W., Haasen, P. (eds) Physical Metallurgy Vol. **1**, 134 North-Holland, Amsterdam (1996)
255. Landa, A., Klepeis, J., Söderlind, P., Naumov, I., Velikokhatnyi, O., Vitos, L., Ruban, A.: J. Phys. Chem. Solids **67**, 2056-2064 (2006)
256. Huang, L., Vitos, L., Kwon, S. K., Johansson, B., Ahuja, R.: Phys. Rev. B **73**, 104203 (2006)
257. Ding, Y., Ahuja, R., Shu, J., Chow, P., Luo, W., Mao, H.-k.: Phys. Rev. Lett. **98**, 085502 (2007)
258. Ashkenazi, J., Docorogna, M., Peter, M., Talmor, Y., Walker, E., Steinemann, S.: Phys. Rev. B **18**, 4120 (1978)
259. Kissavos, A. E., Simak, S. I., Olsson, P., Vitos, L., Abrikosov, I. A.: Comput. Mater. Sci. **35** 1 (2006)
260. Staunton, J., Györffy, B. L., Pindor, A. J., Stocks, G. M. Winter, H.: J. Magn. Magn. Mater. **45**, 15 (1984)
261. Györffy, B. L., Pindor, A. J., Staunton, J., Stocks, G. M., Winter, H.: J. Phys. F: Met. Phys. **15**, 1337 (1985)
262. Hultgren, R., Desai, P. D., Hawkins, D. T., Gleiser, M., Kelley, K. K.: In: Selected Values of Thermodynamic Properties of Binary Alloys. American Society for Metals, Ohio (1973)
263. Vitos, L., Korzhavyi, P. A., Johansson, B.: Phys. Rev. Lett. **96** 117210, (2006)
264. Olsson, P., Abrikosov, I. A., Vitos, L., Wallenius, J.: J. Nucl. Mater. **321**, 84 (2003)
265. Olsson, P., Domain, C., Wallenius, J.: Phys. Rev. B **75**, 014110(12) (2007)
266. Klaver, T. P. C., Drautz, R., Finnis, M. W.: Phys. Rev. B **74**, 094435(11) (2006)
267. Ackland, G. J.: Phys. Rev. Lett. **97**, 015502(4) (2006)
268. Landa, A., Söderlind, P., Ruban, A., Vitos, L., Pourovskii, L. V.: Phys. Rev. B, **70**, 224210 (2004)
269. Olsen, S. J., Gerward, L., Benedict, U., Itie, J.-P.: Physica B and C **133**, 129 (1985)
270. Vohra, Y. K., Akella, J.: Phys. Rev. Lett. **67**, 3563 (1991)
271. Vohra, Y. K., Akella, J.: High Press. Res. **10**, 681 (1992)
272. Gu, G., Vohra, Y. K., Benedict, U., Spirlet, J. C.: Phys. Rev. B **50**, 2751 (1994)
273. Gu, G., Vohra, Y. K., Winand, J. M., Spirlet, J. C.: Scr. Metall. Mater. **32**, 2081 (1995)

274. Söderlind, P., Eriksson, O., Johansson, B., Wills, J. M.: Phys. Rev. B **52**, 13169 (1995)
275. Donohue, J.: In: The Structure of Elements. Wiley, New York, (1974)
276. Vohra, Y. K., Holzapfel, W. B.: High Press. Res. **11**, 223 (1993)
277. Beaudry, B. J., Palmer, P. E.: J. Less-Common Met. **34**, 225 (1974)
278. Bain, E. C.: Trans. AIME **70**, 25 (1924)
279. Söderlind, P., Eriksson, O.: Phys. Rev. B **60**, 9372 (1999)
280. Monnartz, P.: Z. Metallurgie **8**, 161 (1911).
281. Wrangln, G.: An Introduction to Corrosion and Protecting of Metals. Chapman and Hall, New York (1985)
282. Park, E., Hüning, B., Spiegel, M.: Appl. Surf. Sci. **249**, 127 (2005)
283. Lince, J. R., Didziulis, S. V., Shuh, D. K., Durbin, T. D., Yarmoff, J. A.: Surf. Sci. **277**, 43 (1992)
284. Nonas, B., Wildberger, K., Zeller, R., Dederichs P.H.: Phys. Rev. Lett. **80**, 4574 (1998)
285. Suzuki, S., Kosaka, T., Inoue, H., Isshiki, M., Waseda, Y.: Appl. Surf. Sci. **103**, 495 (1996)
286. Geng, W. T.: Phys. Rev. B 68, 233402 (2003)
287. Ruban, A. V., Skriver, H. L., Norskov, J. K.: Phys. Rev. B, 59, 15990 (1999)
288. Pourovskii, L., Ruban, A., Abrikosov, I., Vekilov, Y. K., Johansson, B.: Phys. Rev. B **64** 035421 (2001)
289. Ropo, M., Kokko, K., Vitos, L., Kollár, J.: Phys. Rev. B, **71**, 045411 (2005)
290. Ropo, M., Kokko, K., Vitos, L., Kollár, Johansson, B.: Surf. Sci. **600**, 904-913 (2006)
291. Ropo, M.: Phys. Rev. B **74**, 195401(4) (2006)
292. Wouda, P., Schmid, M., Nieuwenhuys, B., Varga, P.: Surf. Sci. **417**, 292 (1998)
293. Reniers, F.: Surf. Interf. Anal. **23**, 374 (1995)
294. Majumdar, A. K., and Blanckenhagen, P. v.: Phys. Rev. B **29**, 4079 (1984)
295. Ryan, M. P., Williams, D. E., Chater, R. J., Hutton, B. M., McPhail, D. S.: Nature **415**, 770 (2002)
296. Steinsvoll, O., Majkrzak, C. F., Shirane, G. Wicksted, J.: Phys. Rev. Lett. **51**, 300 (1983)
297. Grimvall, G.: Phys. Rev. B **39**, 12300 (1989)
298. Grimvall, G., Hglund, J. Guillermet, A. F.: Phys. Rev. B **47**, 15338 (1993)
299. Grimvall, G.: Physica Scripta **13**, 59 (1976)
300. Oguchi, T., Terakura, K. Hamada, N.: J. Phys. F: Met. Phys. **13**, 145 (1983)
301. Pindor, A. J., Staunton, J., Stocks, G. M. Winter, H.: J. Phys. F: Met. Phys. **13**, 979 (1983)
302. Staunton, J., Györffy, B. L., Stocks, G. M. Wadsworth, J.: J. Phys. F: Met. Phys. 16 1761 (1986)
303. Staunton, J., Györffy, B. L.: Phys. Rev. Lett. **69**, 371 (1992)
304. Pinski, F. J., Staunton, J., Gyrffy, B. L., Johnson, D. D. Stocks, G. M.: Phys. Rev. Lett. **56**, 2096 (1986)
305. Nath, T. K., Sudhakar, N., McNiff, E. J., Majumdar, A. K.: Phys. Rev. B **55**, 12389 (1997)

306. Geng, Y. J., Wang, D. H., Chen, B. Q., Cui, F. Z.: J. Phys. D Appl. Phys. **28**, 226 (1995)
307. Ledbetter, H. M., and Kim, S. A.: J. Mater. Res. **3**, 40 (1988)
308. Ledbetter, H. M.: Physica **128B**, 1 (1985)
309. Ledbetter, H. M.: Met. Sci., 595 (1980)
310. Vitos, L., Korzhavyi, P. A., Johansson, B.: Materials Today, October, 14 (2002)
311. Ledbetter, H. M., Austin, M. W.: J. Mater. Sci. **33**, 3120 (1988)
312. Dyson, D. J., Holmes, B.: J. Iron Steel Inst. May, 469 (1970)
313. Friedel, J.: In: Ziman, J. M. (ed.) The Physics of Metals. Cambridge University Press, p. 494 (1969)
314. Smallman, R. E.: Modern Physical Metallurgy, 4th edition, Butterworths (1985)
315. Rhodes, C. G., Thompson, A. W.: Met. Trans. A **8**A, 1901 (1977)
316. Reick, W., Pohl, M., Padilha, F.: Steel Res. **67**, 253 (1996)
317. Petrov, Yu. N.: Phys. Met. **6**, 735 (1985)
318. Petrov, Yu. N.: Z. Metallkd. **94**, 1012 (2003)
319. Martinez, L. G., Imakuma, K., Padilha, A. F.: Steel Res. **63**, 221 (1992)
320. Brofman, P. J., Ansell, G. S.: Met. Trans. A **9A**, 879 (1978)
321. Nie, X., Zhou, Y., Zhong, L., Wang, D.: J. Mater. Sci. Technol. **2**, 72 (1996)
322. Ferreira, P. J., Müllner, P. A.: Acta Mater. **46**, 4479 (1998)
323. Yakubtsov, I. A., Ariapour, A., Petrovic, D. D.: Acta Mater. **47**, 1271 (1999)
324. Vitos, L., Nilsson, J.-O., Johansson, B.: Acta Mater. **54**, 3821 (2006)
325. Skriver, H. L.: Phys. Rev. B **31**, 1909 (1985)
326. Latanision, R. M., Ruff, A. W. JR.: Metallurg. Trans. **2**, 505 (1971)
327. Olson, G. B., Cohen, M. A.: Metallurg. Trans. A **7**, 1897 (1976)
328. Abrassart, F.: Metallurg. Trans. **4**, 2205 (1973)
329. Oliver, G. L., Perdew, J. P.: Phys. Rev. A **20**, 397 (1979)
330. Perdew, J. P., Ernzerhof, M., Zupan, A., Burke, K.,: J. Chem. Phys. **108**, 1522 (1998)
331. Vitos, L., Johansson, B., Kollár, J., Skriver, H.L.: Phys. Rev. A **61**, 52511 (2000)
332. Becke, A. D.: J. Chem. Phys. **85**, 7184 (1986)
333. Görling, A.: Phys. Rev. Lett. **83**, 5459 (1999)
334. Edmonds, A. R.: In: Wigner, E., Hofstadter, R. (eds) Angular Momentum in Quantum Mechanics. Princeton University Press, Princeton (1957)

Index